The Wave Concept in Electromagnetism and Circuits

Series Editor
Daniel Pasquet

The Wave Concept in Electromagnetism and Circuits

Theory and Applications

Edited by

Henri Baudrand, Mohammed Titaouine,
Nathalie Raveu

ISTE

WILEY

First published 2016 in Great Britain and the United States by ISTE Ltd and John Wiley & Sons, Inc.

ISTE Ltd
27-37 St George's Road
London SW19 4EU
UK

www.iste.co.uk

John Wiley & Sons, Inc.
111 River Street
Hoboken, NJ 07030
USA

www.wiley.com

Library of Congress Control Number: 2016945696

British Library Cataloguing-in-Publication Data
A CIP record for this book is available from the British Library
ISBN 978-1-84821-959-5

Contents

Preface

Summary

The Wave Concept Iterative Procedure (WCIP) was first published in 1995 to model planar circuits and antennas. The first method nearest to the WCIP was harmonic balance (1975) that introduced a successive Fourier Transform and its inverse. The balance between two states, linked by these transforms, is now fundamental to elaborate a model for non-linear electromagnetism problems and high frequency circuits.

After Chapter 1, which introduces the expression and the history of the WCIP, the structures studied are decomposed in elements that link incident and reflected waves, as in the Transmission Line Matrix (TLM) method. The WCIP is quite similar to an integral form of the TLM. One can show that the WCIP is always convergent. After this description of the WCIP and the introduction of its applications in the following chapters, the general principles of the WCIP are presented. They are suitable for the simulation of Meta materials, Substrate Integrated Circuits (SIC), and more generally almost-periodic circuits. As a consequence, dual quantities have to be defined (for instance **E** and **H**, **V** and **I, E** and **J**) and two complementary domains (interior and exterior, spatial and spectral) are presented, as in harmonic balance. The study of nano-transistors is also possible with this method.

In Chapter 2, the WCIP is extended to multilayer structures. New relations between waves are introduced for the transformation between two adjacent interfaces. This formulation is obtained from the impedance matrix of a transmission line, and can be used to model real and ideal polarizer in planar structure. It can be demonstrated that an ideal polarizer interface can be expressed directly in the modal domain, reducing significantly the computational time. One of the applications of a multilayer structure with ideal polarizer is a compact microwave amplifier structure. The WCIP has therefore been modified to characterize the amplifier composed of passive and active elements.

In Chapter 3, Frequency Selective Surfaces (FSS) are widely used as low pass filters, high pass filters, stop band filters, pass band filters, etc. Loaded FSSs allow flexibility in the adjustment of transmission characteristics. However, electronically adjusted FSSs – also called active FSSs are very important in compensating for the errors of fabrication and for the matching of FSS filters for wireless communication systems. The use of short circuits in the dielectric ring FSS ensure the possibility of producing an FSS with a given resonant frequency only by varying the short circuit positions in the ring. When seeking FSS for multiband applications, FSS with concentric rings and multilevel FSSs can be used. The right choice of the FSS circuit can offer the possibility of exploiting the two perpendicular polarisations with distinct multiband. For enhanced band FSS with better filtering characteristics when switching from pass band to stop band, multilevel FSS is used. Inhomogeneous FSS can be accounted for when a need for a specific non extant dielectric substrate is necessary. In integrated circuits the inhomogeneous case corresponds to physical reality. Anisotropic FSS provides filtering behavior with less power loss, thus its study is of great importance. In the quasi periodic FSS, FSS with loaded short circuits, passive elements and active elements are considered as FSS with defects the coupling between all the elements constituting the defected FSS is accounted for.

In Chapter 4, the circuits in substrate integrated circuits (SIC) technology also present the advantage of being integrable with other technologies such as planar circuits or waveguides. To design such circuits, usual numerical techniques may be inappropriate: the Finite Element Method implies important memory storage while the Integral Method is not able to solve dielectric vias integration. The WCIP is therefore an efficient method for characterizing SIC when vias are placed on a periodic lattice in metallic and dielectric cases. Several example circuits are analysed in this chapter and the computation time is very low.

In Chapter 5, the convergence of the WCIP is proved. Convergence is ensured as the eigenvalues of the WCIP operators are bounded by one; however the convergence may be reached very slowly. This method, well adapted to several configurations of circuits, has been improved in terms of its computation time by two ways: physical parameter modification and mathematic development. Physical parameters, such as the reference impedance in expression of waves and metallic loss insertion, help the convergence; however, performance is not drastically changed. Mathematic development of other numerical methods is well-known such as Krylov solution instead of point fixed solution, physical pre-conditioners etc. They are implemented and proved to be efficient. Thanks to these improvements the computation time obtained in Chapter 5 has been improved by almost 95% compared to the usual solution for same tolerance on results.

The electromagnetic waves scattered by metallic or inhomogeneous objects is of interest in several areas, such as radar and CEM. In Chapter 6, the formulation of the WCIP in the cylindrical coordinate system is used to study scattering problems and the penetration of electromagnetic waves into multilayer cylindrical structures. In this chapter, the operators inside and outside a cylinder, and the operator ensuring the wave transmission between two coaxial cylinders defined in spectral domain, are used to calculate the incident and reflected waves at each interface of the multilayered cylindrical structure. Finally, the WCIP method is extended to model the scattering of a structure of arbitrary shape in an open environment. The development of the diffraction operator is expanded in cylindrical modal basis in the spectral domain. In fact, the basis functions should be created in the most appropriate coordinate system, suitable to the local geometry of the arbitrary diffraction structure. For a planar circuit, the scattering problem is easily solved by the WCIP method in the Cartesian coordinate system. However, the Cartesian coordinate system becomes inefficient when dealing with non-regularly shaped bodies. A cylindrical coordinate formulation is presented in this section, as a solution to extend the WCIP to scattering of structures of arbitrary shapes.

These six chapters show the large range of possible applications in electromagnetism that are solved with the WCIP. The main advantage of the WCIP lies in the fact the convergence is ensured. This method has also been generalized to a large domain of applications that are not presented in this book, such as almost-periodic structures, non-linear media and non-linear impedances, exploration of time domain with multi-Fourier transform and 3-D WCIP deduced directly from Maxwell Equations.

Herri BAUDRAND,
Mohammed TITAOUINE and Nathalie RAVEU
June 2016

1

General Principles of the Wave Concept Iterative Process

1.1. Introduction

The iterative method, which uses a wave network, is an integrated method and is not based upon electric and magnetic fields, as are, for example, Electrical Field Integral Equation (EFIE), Magnetic Field Integral Equation (MFIE), or more generally the method of moments or a combination of both fields. These are likened to the amplitudes of transverse waves, both diffracting around obstacles and those in space, termed "free space", owing to the presence of evanescent fields. However, while the method of moments appeals to so-called admittance or impedance operators, within the wave iterative method (Wave Concept Iterative Process (WCIP)), the diffraction operators are restricted, thus leading to the convergence of all iterative processes based upon this particular formalism [BAU 99].

It may be noted that, with the method of moments, the solution to the problem often entails using a restriction in the given field so as to define trial functions that constitute the basis for given solutions. This often leads to both analytical and numerical problems. In the WCIP method, field conditions are simply described on the basis of pixels which make up the entire sphere.

Chapter written by Henri BAUDRAND, Med Karim AZIZI and Mohammed TITAOUINE.

Moreover, the iterative process has a significant resemblance to that used within harmonic equilibrium [KER 75]. Within this latter process the nonlinear component behaves in a way that is described in relation to time, while the rest of the circuit is described within the frequency sphere. The operator thus functions diagonally at given frequencies. With each iteration, we therefore proceed with a Fourier transform (using a time–frequency basis) so as to approach the detailed composition of boundary conditions at the shutdown level. Moreover, when writing equations in terms of components studied over time, an inverse Fourier transform (based upon frequency–time) is used.

The WCIP approach is closely related. By simply replacing time by a coordinate and the frequency by a "spatial frequency", the operation reverts to one within the spectral sphere. Outside of the Transverse Lines Matrix (TLM) method, which also necessitates the wave concept [KRU 94], the WCIP is based upon the systematic iteration between both incident and reflected waves. The approach used in the paragraphs below is as follows: select a wave definition which is consistent with pre-existing cases, in particular within waveguides, and ensure that it has a fundamental physical significance. The iterative process will then be described in the context of several types of problems, in particular quasi-periodic structures.

The objectives of this chapter are to first set out the WCIP, showing its potential for circuit modeling, antennae and quasi-optical devices within stratified environments [BAU 99, AZI 95, AZI 96, WAN 05, RAV 04, TIT 09]. There are two advantages to this method. Firstly, the iterative process is always convergent (excepting the frequency resonating from a mechanism such as that one which is also relevant to other digital methods). Secondly, by the description of all surfaces through the use of pixels, it is not necessary to use a network describing the part of the surface corresponding to a metallic coating (or indeed to the dielectric dual), as falls within the sphere of the method of moments.

In the second part of this chapter, the WCIP is outlined. The principles of the WCIP are adhered to. Through the use of combined equations, one is expressed in the spatial sphere and the other in the spectral sphere (also called the modal sphere). The solution is obtained by achieving equilibrium between these two spheres. The description of a given mechanism is not set by rectangular pixels but by cells restricted by periodic barriers, each containing periodic non-configured sources. The sources are described within the spectral sphere (defined by periodic barriers). They are called “auxiliary sources”, as they need to be substituted by impedances alone or indeed other sources or impedances, the latter being defined by each cell. Hence the use of the term spatial domain and the adjective quasi-periodic being applied to the system. This concept makes it possible to study a large variety of systems, in particular Substrate Integrated Circuits (SIC), which have been successfully developed for several years, as shown by the results from a number of examples.

The last part of the chapter provides a gateway to other interesting applications in the field of non-homogeneous meta-materials, in the sense that both sources and obstacles are integrated in a unique model, thus avoiding the use of the approximation of equivalent environments. Up until now (except for using three-dimensional simulators such as EF and Finite-Difference Time-Domain (FDTD)) homogeneous meta-materials allow us to establish their equivalent index. The link between a material and a given mechanism (for example, plate antennae) presents difficulties and cannot be approached from any perspective other than a comprehensive analysis.

Finally, we provide an overview of another WCIP-based field; the study of quasi-periodic circuits with identified components. There are many applications for these types of structures, filters, amplifiers, percolation problems and quasi-optic planar sources.

1.2. The iterative wave method

The integral form of waves came to be explained during the 1990s, and was applied to planar circuits and to antennae [BAU 99, AZI 95, AZI 96, WAN 05, RAV 04, TIT 09]. The wave concept principle is as follows:

– The electromagnetic issue may be expressed by the relationship between the two environments. The first is known as the spectral sphere or the external environment. The second is a set of surfaces which are defined by the boundary conditions at each point (termed the spatial domain). An Ao source in the spatial sphere sends a wave with an Ao amplitude towards a vacuum of free space. This wave is partly reflected (by the reaction of the operator *Γ*) and provides a wave *B*. The latter is, in its turn, reflected within the spatial sphere (the Operator *S*) giving us the wave *A*.

– The *Γ* operator is diagonal within the spectral sphere. It represents the homogeneous environment and its interaction with electromagnetic waves [BAU 99]. The operator *S* describes the boundary conditions of the interface. It is expressed within the spatial sphere. The Fourier transform and its converse, the inverse Fourier transform, ensure the passage between both spheres. The relationships between incident and reflective waves are written as shown in [1.1] and [1.2].

$$B = \Gamma A \qquad [1.1]$$

$$A = SB + A_0 \qquad [1.2]$$

With the first iteration, the spatial sphere equation should be expressed simply as Ao ($B = 0$). *B* now appears with the operator *Γ* ($B = \Gamma A$). The equation [1.2] is applied so as to obtain the new value of *A* placed within [1.1], resulting in the new *B* value. This iterative process consists in successively applying equations [1.1] and [1.2], until convergence occurs (Figure 1.1).

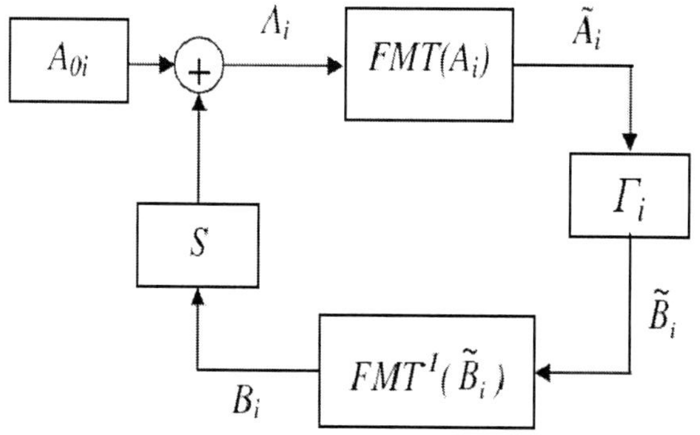

Figure 1.1. *Iterative wave diagram*

For a planar circuit within a rectangular housing, the operator Γ is diagonal across modes TE and TM, as are both the rapid transformation in methods Fast mode transform (FMT) and the opposing linking equations [1.1] and [1.2], this occurs because the operator S is diagonal in the spatial domain.

1.3. General definition of waves

The general definition of waves must meet certain conditions:

The existence of a division of the overall sphere into two sub-spheres: the internal sphere or the spatial sphere (these are flat interfaces or localized elements, indeed centers of boundary conditions within integrated methods). The second sphere is the external sphere (or spectral sphere). This sphere is most often described on the basis of the unique functions of the Helmholtz operator, which stems from Maxwell's equations. To develop this method, we need to define two dual variables such as Current–Voltage, Electric field–Magnetic field, Current density (density or surface)–Electric field, and Voltage–Load density or Voltage–Load. All of the possibilities are shown in Table 1.1. E and J may be taken as two dual variables. J is not necessarily a current-related density, but encompasses all magnitudes which are defined in Table 1.1. J may also be related to current volume density. One would thus write it as Jv to avoid confusion with the magnetic field rotated by 90°(H^n). Wave amplitudes A and B are thus defined (it may be observed that A and B may be scalars or vectors):

$$\vec{A} = \frac{1}{2\sqrt{Z_0}}\left(\vec{E} + Z_0\vec{J}\right) \qquad [1.3]$$

$$\vec{B} = \frac{1}{2\sqrt{Z_0}}\left(\vec{E} - Z_0\vec{J}\right) \qquad [1.4]$$

1.4. Application to planar circuits

The first and most frequent wave representation is where dual variables are tangential components of fields on a surface S (Figure 1.2) adjusted to account for the magnetic field by H^n (the vector product is chosen rather than the magnetic field, simply for reasons of homogeneity). This representation is highly useful when dealing with circuits and planar antennae, frequency-selective surfaces, diffraction issues and systems involving cylindrical or spherical coordinates.

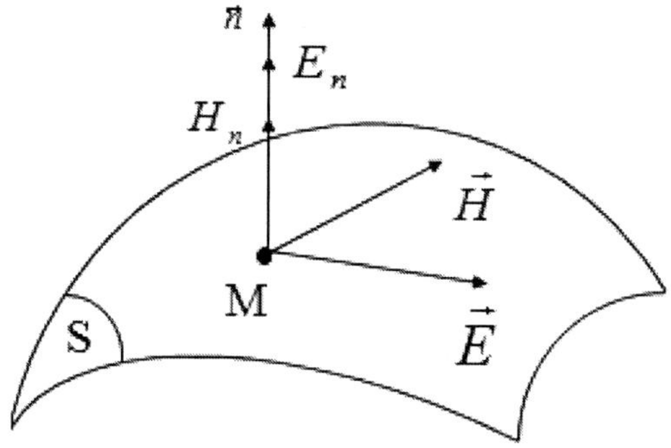

Figure 1.2. *Features of half-spaces*

1.5. Applications to quasi-periodic structures

Other types of new applications have been introduced as so-called "quasi-periodic structures". These use other wave definitions (see Table 1.1) which have applications in planar reflectors, SICs, meta-materials and photonic jets.

Dual variables	*Definition domain*	*The product of dual variables*
General case $\vec{E}\,;\vec{J} = \vec{H} * \vec{n}$	TM0 mode: plane xOy	$\vec{E}.\vec{J}^{*} = \left(\vec{E} \wedge \vec{H}^{*}\right)\vec{n}$
$\vec{E}\,;\vec{J}_s = \vec{H}_{2T} - \vec{H}_{1T}$	Surface with surface current $\vec{J}_s$	$\vec{E}.\vec{J}_s^{*}$ Surface power density
V; I	Lumped element circuits or TEM mode	$V \times I^{*}$ Complete power
V; ρ	Volume with load ρ	ρV Electro-static energy

Table 1.1. *Possible different dual variables*

1.6. Circuits with localized components

The traditional iterative wave method involves breaking down an electromagnetic problem into two parts [BOZ 09] as follows. The propagation equation aspect within a vacuum is dealt with in its entirety, and therefore translates as a relationship with the boundaries across sphere *D*, then with the boundary conditions running across sphere *D*. It is then necessary to have dual magnitudes linked together in a vacuum and at the boundaries, by linear operators, through a proportionality relationship (which is internal to *D*) and an integral relationship (which is external to *D*).

Figure 1.3 shows the unidimensional structure which is made up of several cells, each enclosed by periodic walls. This structure is periodic, except at source level.

$$E_2 = E_1 e^{j\alpha};\ E_3 = E_2 e^{j\alpha};\ E_4 = E_3 e^{j\alpha};\ E_5 = E_4 e^{j\alpha}$$

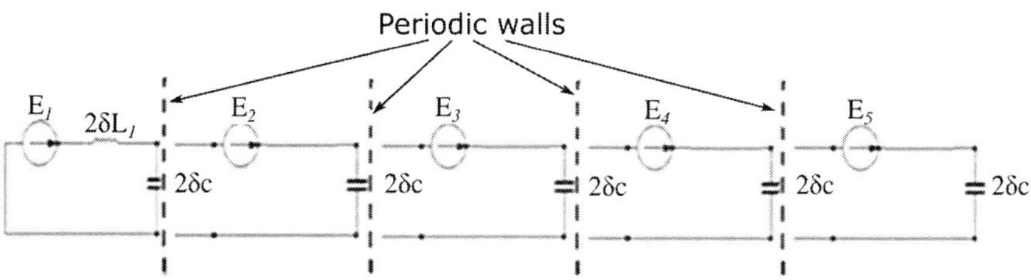

Figure 1.3. *Quasi-periodic univariate structure*

1.7. General principles of quasi-periodic circuits

Within a periodic cell-based structure, each cell is assumed to have a source. Any default at the cell level may be represented by replacing the source of the latter by impedances [WAH 92]. The structure then becomes quasi-periodic. This approach is well adapted to the iterative wave method, which acts as a balance between spectral (exterior) and spatial (internal or source) sphere, since it describes periodic geometry (external) with small spatial irregularities (the so-called sphere of sources).

One example which is often studied relates to Substrate Integrated Circuits (SICs) within an elementary cell made up of a metal or dielectric rod, which is plated by periodic walls (Figure 1.4) [BAU 09].

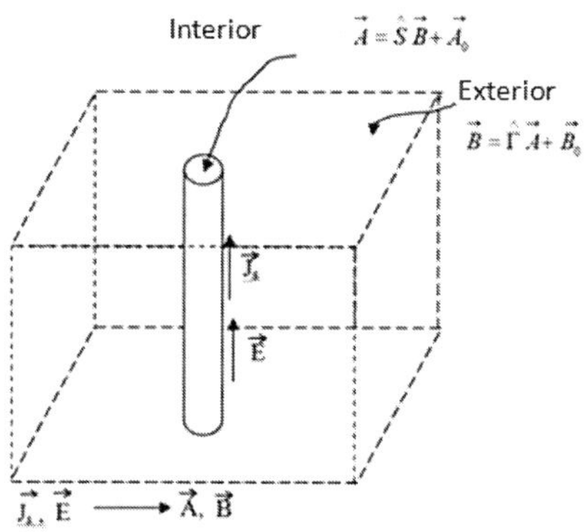

Figure 1.4. *The elementary cell*

1.8. The significance of using auxiliary sources

Consider an electromagnetic environment (Figure 1.5) comprised of one (or several) components whose behavior is described in the system V-I, E-J or even E-Js.

This element may be replaced by a source, the shutdown for which is formed by the electromagnetic environment.

The closed matrix may be established across two given sources; E_0 being the device power source and E_A being the auxiliary source.

As Figure 1.5 shows, this component may be replaced with a source (stage 1), the shutdown for which may be constituted by this component (stage 2).

The calculation produces impedance (and the potential source) in view of this source, from the electromagnetic environment. This will operate within the spectral domain.

Stage (2) involves stating that the source of stage (1) (the arbitrary source) is shutdown on the impedance of the output circuit.

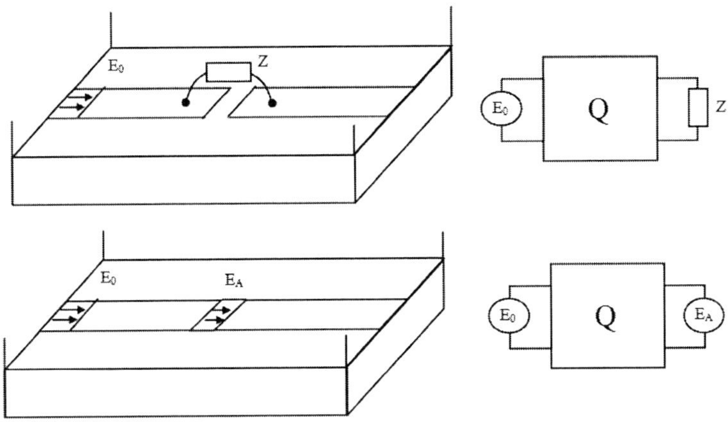

Figure 1.5. *Use of an auxiliary source*

1.8.1. *Description of the environment*

This consists of a system *Q*, both fed by a source S_0 and closed by an impedance *Z*. In general *Q* constitutes the center of the electromagnetic field E.M. To examine this system, we can break it down into two parts, each separate and distinct from the other and fed by a source known as the auxiliary source. System (I) describes the behavior of the impedance *Z*. System (II) describes the main source within its environment; E.M.*Z* can generally be defined at any point, and forms the spatial sphere. On the other hand, *Q* is often described within the environment E.M. *Q* is defined by its impedance or diffraction matrix. It is necessary to resort to the spectral sphere. Using this method, the calculation of impedance from the angle of S_0 is not achieved directly but it is first necessary to calculate a quadrupole with the help of an auxiliary source. This will subsequently be replaced by the localized impedance within the actual issue (allowing for its potential source). This latter operation is known as an operation in the spectral domain.

1.9. Unidimensional circuits

The configuration in Figure 1.6 shows the unidimensional structure, which is made up of several cells.

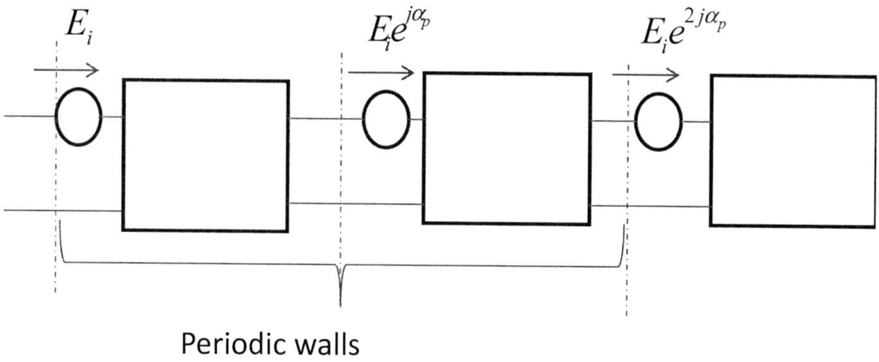

Figure 1.6. *Quasi-periodic unidimensional circuits in the state αp*

The term unidimensional circuits refers to the balanced elementary cell (Figure 1.7) plated with periodic walls.

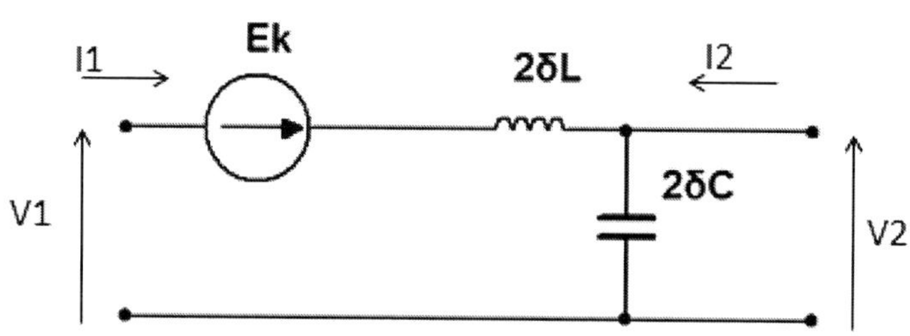

Figure 1.7. *The elementary cell*

It is first necessary to calculate the self-inductance values and the capacity for a 3 Ghz frequency. For a cell with a length δl, the capacity is $2\delta C$ and the self-inductance is $2\delta L$, hence by unit of length the formulae:

$$\delta C = 0.67.10^{-10}\ {}^{*}1.25.10^{-3} = 0.84.10^{-13}\,F$$

$$\delta L = 1.67.10^{-7}\ {}^{*}1.25.10^{-3} = 2.09.10^{-10}\,H$$

$$\begin{array}{l} \delta L \textit{ represents inductance by unit of length} \\ \delta C \textit{ represents capacity by unit of length} \end{array} \qquad [1.5]$$

The study of the structure of an artificial line made up of 100 cells at a frequency F will be completed for a given cell length $\delta l = \lambda / 40 = 2.5\ mm$. Upon isolation of the elementary cell within the unidimensional structure (Figure 1.7), all of the work which will be completed upon this cell then takes account of the periodicity walls surrounding it. This amounts to a saving both in terms of the time taken to complete the calculation, and also precision. Within the state α we must calculate $Z\alpha$ and three relationships thus arise: by taking the elementary cell in (Figure 1.7), Kirchoff's mesh and junction rules (respectively relating to voltage and current) allow us to express the following formulae:

$$V_1 + jl\omega I_1 + E_e + jl\omega I_1 = V_2 \qquad [1.6]$$

$$I_2 = -I_1 e^{j\alpha} \qquad [1.7]$$

$$I_1 e^{j\alpha} + I_1 + j\omega c V_1 e^{j\alpha} = 0 \qquad [1.8]$$

The transition from one cell to another takes place through a phase difference α, each cell being limited by periodic walls, with periodic conditions implying a phase difference between [1.1] and [1.2]. We thus get the formula:

$$V_2 = V_1 e^{j\alpha}_{-} \qquad [1.9]$$

Thus, $V_1 + jl\omega I_1 + E_e + jl\omega I_1 = V_1 e^{j\alpha}$ [1.10]

The junction rule provides us with the equation:

$$I_1 - j\omega c V_2 = -I_2 \qquad [1.11]$$

Two of the last equations [1.7] and [1.8] of this sequence give us:

$$(j\omega c V_1 - I_1) e^{j\alpha} + I_1 + j\omega c V_1 e^{j\alpha} = 0 \qquad [1.12]$$

Hence:

$$V_1 = \frac{e^{j\alpha} - 1}{2 j\omega c e^{j\alpha}} I_1 \qquad [1.13]$$

By substituting the second equation for the first, we arrive at:

$$E = -2jl\omega I_1 + \frac{(e^{j\alpha} - 1)}{2j\omega c e^{j\alpha}} I_1 \qquad [1.14]$$

From this impedance, we deduce the result Γ_α and substitute α (a phase difference between one cell and another) by its value $\alpha = 2\pi p / N$.

Hence, the value known as spectral Γ:

$$\Gamma = \frac{Z_\alpha - Z_0}{Z_\alpha + Z_0} \qquad [1.15]$$

For the spatial sphere, we make the structure anti-symmetric (Figure 1.8). D is now equal to double the length of the line. To achieve a line of three cells requires an inductance $2\delta L$ and a capacitance of $2\delta C$, each having a dimension dl according to the wavelength λ commensurate with given requirements.

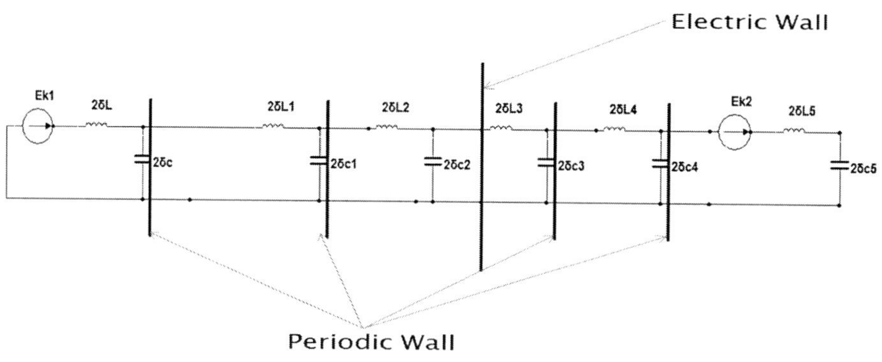

Figure 1.8. *Dividing up the cell configuration (three cells including periodic walls)*

In order to calculate S within the spatial sphere, we have to return to the initial diagram to calculate the parameters of the matrix S at source level. The transmission line acknowledges as the entry impedance Z_0 and a single source in terms of the first cell E_{k1}. For the remainder of the cells, the sources are short-circuited.

This structure allows us to express the formula:

$$E = \begin{pmatrix} E_{k1} \\ 0 \\ 0 \\ 0 \\ 0 \\ 0 \end{pmatrix} + \begin{pmatrix} -Z_0 I_{N/2} \\ 0 \\ 0 \\ 0 \\ 0 \\ 0 \end{pmatrix} \quad [1.16]$$

E represents the electric field.

This occurs by using both equations governing the iterative method and by substituting E in these equations below:

$$\begin{cases} A = \dfrac{1}{2\sqrt{Z_0}}(E + Z_0 \mathrm{J}) \\ \mathrm{B} = \dfrac{1}{2\sqrt{Z_0}}(E - Z_0 J) \end{cases} \quad [1.17]$$

In terms of waves, we thus get:

$$B = \begin{pmatrix} E_{k1}/2\sqrt{Z_0} \\ 0 \\ 0 \\ 0 \\ 0 \\ 0 \end{pmatrix} + \begin{pmatrix} 0 & 0 & 0 & 0 & 0 & 0 \\ 0 & -1 & 0 & 0 & 0 & 0 \\ 0 & 0 & -1 & 0 & 0 & 0 \\ 0 & 0 & 0 & -1 & 0 & 0 \\ 0 & 0 & 0 & 0 & -1 & 0 \\ 0 & 0 & 0 & 0 & 0 & -1 \end{pmatrix} \begin{pmatrix} A_{-N/2} \\ . \\ . \\ . \\ . \\ A_{N/2-1} \end{pmatrix} \quad [1.18]$$

This formula represents waves within the spatial domain. By doing so, we thus acknowledge equation [1.2] of the iterative method, B = SA+ B_0, as well as equation [1.15], allowing the determination of impedance within the spectral domain. Having calculated S and Γ, the current density I placed upon each cell may be determined.

1.10. Application: transmission line

Knowing the spectral Γ and the spatial S with the main sources E_1 the iterative process may be initiated and the values of I on each of the actual cells instead of auxiliary sources may be determined.

Convergence is reached after 2,800 iterations. This makes it possible to calculate the current density passing through each cell, that is to say a cell length of $\delta l = \lambda/10$.

By inserting the electric wall at the point of the 50th cell, the structure was split into two and the current density at its maximum, which allowed us to find the wavelength for 100 cells. Figure 1.10 shows current distribution for another cell length of the order of $\delta l = \lambda/50$.

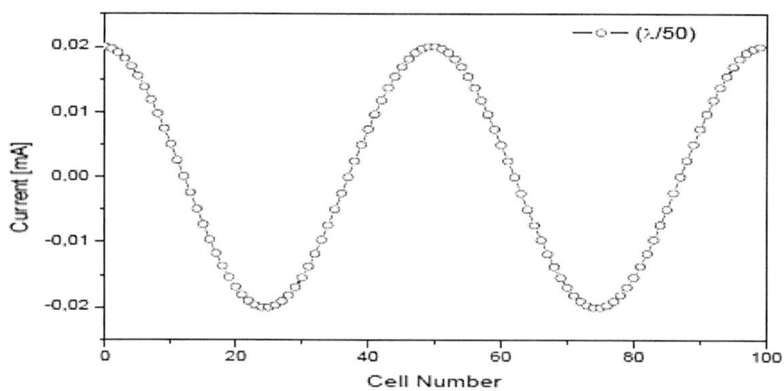

Figure 1.9. *Current according to cell number for a length λ/50*

1.11. Comparison of current density for different cell lengths

Let us now take, by way of length, the artificial line Lg = 30 cm. A simulation of this structure was completed. This thus allowed us to see the current behavior on each cell, for different cell lengths. A superimposition of three curves for current density (Figure 1.10) has been shown to see what length is most appropriate for this function.

So as to permanently fix concepts upon the choice of cell length, another length of the artificial line of the order of 15 cm was selected. This trial allows us to note that the length at which current density is stable is

$\delta l = \lambda/40$. The curve below (Figure 1.11) shows the theoretical current density component. Since it is a short circuit line, there are no losses, giving us theoretical pure impedance.

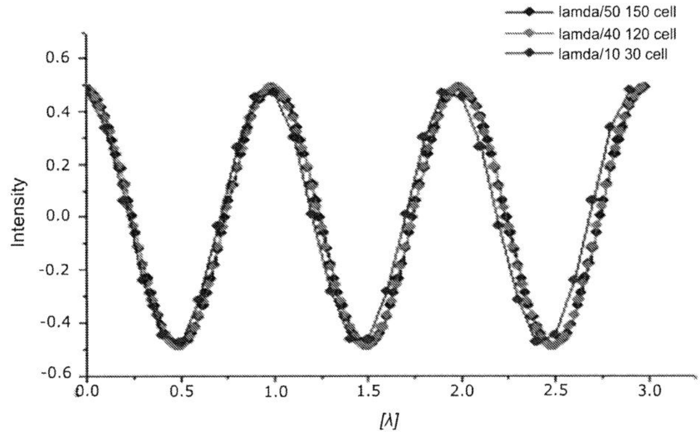

Figure 1.10. *Comparison of current density for different cell lengths for a length of 30 cm. For a color version of this figure, see www.iste.co.uk/baudrand/waves.zip*

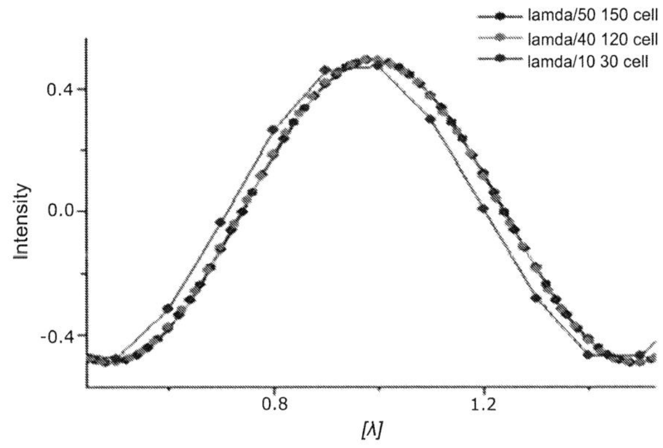

Figure 1.11. *Comparison of current density for different cell lengths measured according to wavelength λ, For a color version of this figure, see www.iste.co.uk/baudrand/waves.zip*

In the work which follows the cell length will be $\delta l = \lambda/40$, because current stability upon a cell is guaranteed for this length.

1.12. Bi-dimensional circuits

The periodic bi-dimensional structure which is made up of the cells (on the basis of N*M) will be studied. Owing to its periodicity, the elementary cell is isolated and the entire study is made up of the latter. The cell is plated by periodic walls. The transition between one cell and another takes place through two clearly established bi-directional phase differences.

1.13. Two-source bi-dimensional circuits

Figure 1.12 shows an elementary cell for detailed study.

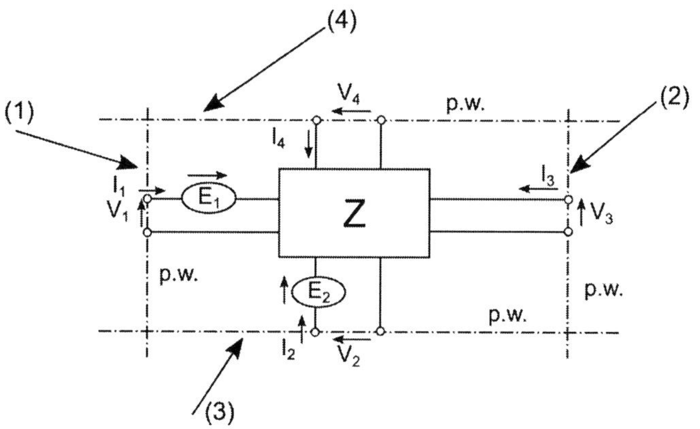

Figure 1.12. *Elementary cell*

Periodic conditions imply a phase difference between [1.1] and [1.2] and also between [1.3] and [1.4].

$$\begin{cases} V_2 = V_1 e^{j\alpha} \\ I_2 = -I_1 e^{j\alpha} \\ V_4 = V_3 e^{j\beta} \\ I_4 = -I_3 e^{j\beta} \end{cases} \qquad [1.19]$$

α and β are arbitrary. They occur because of the form, within a housing with the dimensions $D*D$, appreciating that the cell is of a dimension of d′*d:

$$\alpha(m) = 2\pi m\frac{d}{D} \text{ and } \beta(n) = 2\pi n\frac{d}{D} \qquad [1.20]$$

Moreover:

$$\begin{vmatrix} V_1 \\ V_2 \\ V_3 \\ V_4 \end{vmatrix} = \begin{vmatrix} Z_{11} & Z_{12} \\ Z_{21} & Z_{22} \end{vmatrix} \begin{vmatrix} I_1 \\ I_2 \\ I_3 \\ I_4 \end{vmatrix} - \begin{vmatrix} E_1 \\ 0 \\ E_3 \\ 0 \end{vmatrix} \qquad [1.21]$$

There are eight equations with eight unknowns:

$$\begin{vmatrix} V_1 \\ V_2 \end{vmatrix} = \begin{vmatrix} 1 \\ e^{j\alpha} \end{vmatrix} V_1 \ ; \ \begin{vmatrix} I_1 \\ I_2 \end{vmatrix} = \begin{vmatrix} 1 \\ -e^{j\alpha} \end{vmatrix} I_1 \qquad [1.22]$$

$$\begin{vmatrix} V_3 \\ V_4 \end{vmatrix} = \begin{vmatrix} 1 \\ e^{j\beta} \end{vmatrix} V_3 \ ; \ \begin{vmatrix} I_3 \\ I_4 \end{vmatrix} = \begin{vmatrix} 1 \\ -e^{j\beta} \end{vmatrix} I_3 \qquad [1.23]$$

By using the formula:

$$\begin{vmatrix} 1 \\ e^{j\alpha} \end{vmatrix} V_1 = Z_{11} \begin{vmatrix} 1 \\ -e^{j\alpha} \end{vmatrix} I_1 + Z_{12} \begin{vmatrix} 1 \\ -e^{j\beta} \end{vmatrix} I_3 - \begin{vmatrix} E_1 \\ 0 \end{vmatrix}$$

and [1.24]

$$\begin{vmatrix} 1 \\ e^{j\beta} \end{vmatrix} V_3 = Z_{21} \begin{vmatrix} 1 \\ -e^{j\alpha} \end{vmatrix} I_1 + Z_{22} \begin{vmatrix} 1 \\ -e^{j\beta} \end{vmatrix} I_3 - \begin{vmatrix} E_3 \\ 0 \end{vmatrix}$$

By multiplying [1.22] by $\left|e^{j\alpha}, -1\right|$ we arrive at:

$$0 = \left|e^{j\alpha}, -1\right| Z_{11} \begin{vmatrix} 1 \\ -e^{j\alpha} \end{vmatrix} I_1 + \left|e^{j\alpha}, -1\right| Z_{12} \begin{vmatrix} 1 \\ -e^{j\beta} \end{vmatrix} I_3 - E_1 e^{j\alpha} \qquad [1.25]$$

By multiplying [1.23] by $\left|e^{j\beta},-1\right|$ we arrive at:

$$0= \left|e^{j\beta},-1\right|Z_{21}\begin{vmatrix}1\\-e^{j\alpha}\end{vmatrix}I_1+\left|e^{j\alpha},-1\right|Z_{22}\begin{vmatrix}1\\-e^{j\beta}\end{vmatrix}I_3-E_3e^{j\beta} \qquad [1.26]$$

This brings us back to writing:

$$\begin{cases}E_1(\alpha\beta)=Z_{\alpha\beta}^{(1,1)}\,I_1(\alpha\beta)+Z_{\alpha\beta}^{(1,3)}I_3(\alpha\beta)\\E_3(\alpha\beta)=Z_{\alpha\beta}^{(3,1)}\,I_1(\alpha\beta)+Z_{\alpha\beta}^{(3,3)}I_3(\alpha\beta)\end{cases} \qquad [1.27]$$

These relationships are described in the spectral domain, which is characterized by α and β.

The condensed form is written as:

$$\tilde{I}_{\alpha\beta}=\begin{vmatrix}I_1(\alpha\beta)\\I_3(\alpha\beta)\end{vmatrix}, \quad \tilde{E}_{\alpha\beta}=\begin{vmatrix}E_1(\alpha\beta)\\E_3(\alpha\beta)\end{vmatrix}$$

The equation [1.28] is thus written as:

$$\tilde{I}_{\alpha\beta}=Y_{\alpha\beta}\tilde{E}_{\alpha\beta} \qquad [1.28]$$

Hence, it is possible to work out $\Gamma_{\alpha\beta}$.

The spatial domain corresponds to internal relationships with each source. For this, the transition by Fast Fourier transform analysis to the spectral domain is necessary.

That is to say:

$$\tilde{I}_{mn}=Y_{mn}\tilde{E}_{mn} \text{ with } \alpha(\mathrm{m})=\frac{2\pi m}{N},\beta=\frac{2\pi n}{N} \text{ and } \mathrm{N}=\frac{D}{d} \qquad [1.29]$$

The cells will be numbered by the integers *k*,*l*. The source phase in $k = l = 0$. The phase difference of the source *k*,*l* will be:

$$\begin{cases}k\alpha \ \text{ according to x}\\ l\beta \ \text{ according to y}\end{cases}$$

As there is an α, β phase difference during the transition from one cell to another, we may therefore write:

$$I_{kl} = \sum_{m,n} \tilde{I}_{mn} \underbrace{e^{j\frac{2\pi mk}{N}}}_{\text{phase difference with a displacement of } x} \times \underbrace{e^{j\frac{2\pi nl}{N}}}_{\text{phase difference with a displacement of y}} \qquad [1.30]$$

In these conditions, $\tilde{I}_{mn}$ no longer represents the intensity of sources, but this intensity multiplied by N.

$$I_{kl=}\frac{1}{N}\sum_{m,n}\left(N\tilde{I}_{mn}\right)e^{j\frac{2\pi mk}{N}}\ e^{j\frac{2\pi nl}{N}} \qquad [1.31]$$

The inverse formula for [1.31] is written as:

$$N\tilde{I}_{mn=}\frac{1}{N}\sum_{k,l}I_{kl}e^{-j\frac{2\pi mk}{N}}\ e^{-j\frac{2\pi nl}{N}} \qquad [1.32]$$

The relationship in the spatial domain thus becomes:

$$I_{kl} = \underbrace{Y'_{kl}}_{\text{resistance at source level (internal impedance)}} \mathrm{E}_{kl} + \underbrace{\mathrm{I}^0_{kl}}_{\text{potential source}} \qquad [1.33]$$

The spatial relationship

$$\tilde{I}_{mn} = Y_{mn}\tilde{E}_{mn} \qquad [1.34]$$

Hence, the formula in waves being:

$B = SA + B_0$ (Spatial)

$\mathrm{A}=\hat{\Gamma}B$ (Spectral)

The iterative process breaks down the problem into two parts, one in the spatial domain and the other in the spectral sphere.

Returning to the elementary cell (Figure 1.13) surrounded by periodic walls, there are four elementary cell equations which arise:

$$\begin{cases} V_3 e^{j\beta} = V_1 e^{j\alpha} \\ I_1 e^{j\alpha} + I_3 e^{j\beta} = I_1 + I_3 \\ V_1 e^{j\alpha} - V_1 = E_1 - rI_1 \\ V_3 e^{j\beta} - V_3 = E_3 - rI_3 \end{cases} \qquad [1.35]$$

In a given case, if the Z diagonal $= r$, we find a conventional circuit:

$$I_1\left(e^{j\alpha} - 1\right) = \left(1 - e^{j\beta}\right) I_3$$

By transposing [1.1] and [1.2] within [1.3] and [1.4] into the relationship given at [1.35] we arrive at:

$$V_1\left(e^{j\alpha} - 1\right) = E_1 - zI_1 \qquad [1.3']$$

$$V_1 e^{j(\alpha-\beta)}\left(e^{j\beta} - 1\right) = E_3 - z\frac{e^{j\alpha} - 1}{1 - e^{j\beta}} I_1 \qquad [1.4']$$

Transposing [1.3′] into [1.4′] gives us:

$$(E_1 - rI_1)e^{j(\alpha-\beta)} \frac{e^{j\beta} - 1}{e^{j\alpha} - 1} = E_3 + \frac{e^{j\alpha} - 1}{e^{j\beta} - 1} I_1 \qquad [1.36]$$

or

$$E_1 e^{j(\alpha-\beta)} \frac{e^{j\beta} - 1}{e^{j\alpha} - 1} - E_3 = I_1\left(\frac{e^{j\alpha} - 1}{e^{j\beta} - 1} + e^{j(\alpha-\beta)} \frac{e^{j\beta} - 1}{e^{j\alpha} - 1}\right) \qquad [1.37]$$

that is to say:

$$a = e^{j(\alpha-\beta)}$$
$$b = \frac{e^{j\beta} - 1}{e^{j\alpha} - 1} \qquad [1.38]$$

with: $I_1 = -bI_3$

The previous relationship is written as:

$$abE_1 - E_3 = I_1\left(\frac{1}{b} + ab\right) = -I_3\left(1 + ab^2\right) \quad [1.39]$$

Hence, the admittance matrix:

$$\begin{vmatrix} I_1 \\ I_3 \end{vmatrix} = \begin{vmatrix} \dfrac{ab^2}{1+ab^2} & \dfrac{-b}{1+ab^2} \\ -\dfrac{ab}{1+ab^2} & \dfrac{1}{1+ab^2} \end{vmatrix} \begin{vmatrix} E_1 \\ E_3 \end{vmatrix} \quad [1.40]$$

Although it is surprising, the matrix is non-reciprocal. It would be sufficient to change the origin of the I_3 phases. To achieve this:

In setting out $I_3 = aI_3^{'}$, the relationship between current density and the magnetic field, this formula becomes:

$$\begin{vmatrix} I_1 \\ I_3' \end{vmatrix} = \begin{vmatrix} \dfrac{ab^2}{1+ab^2} & \dfrac{-b}{1+ab^2} \\ -\dfrac{b}{1+ab^2} & \dfrac{1}{a\left(1+ab^2\right)} \end{vmatrix} \begin{vmatrix} E_1 \\ E_3 \end{vmatrix} \quad [1.41]$$

$$\begin{vmatrix} \dfrac{ab^2}{1+ab^2} & \dfrac{-b}{1+ab^2} \\ -\dfrac{b}{1+ab^2} & \dfrac{1}{a\left(1+ab^2\right)} \end{vmatrix}$$ represents the admittance matrix $Y\alpha\beta$ [1.42]

From [1.42], the matrix $\hat{\Gamma}_{\alpha\beta}$ may be calculated by the relationship:

$$\hat{\Gamma}_{\alpha\beta} = \frac{1 - rY_{\alpha\beta}}{1 + rY_{\alpha\beta}}$$

With respect to internal relationships:

$$B = SA + B_0$$

$$S = \begin{cases} -1 \; short\ circuit \\ 1 \; open\ circuit \end{cases}$$

1.14. Three-source bi-dimensional circuits

These are as per the following description of the elementary cell with a transmission line (Figure 1.14) having the length δl, which is made up of three auxiliary sources E_1, E_3 and E_5 following the three respective directions *X*, *Y* and *Z* and the resistance *R*. The source is fed into point *Z* and polarized at point *Y*.

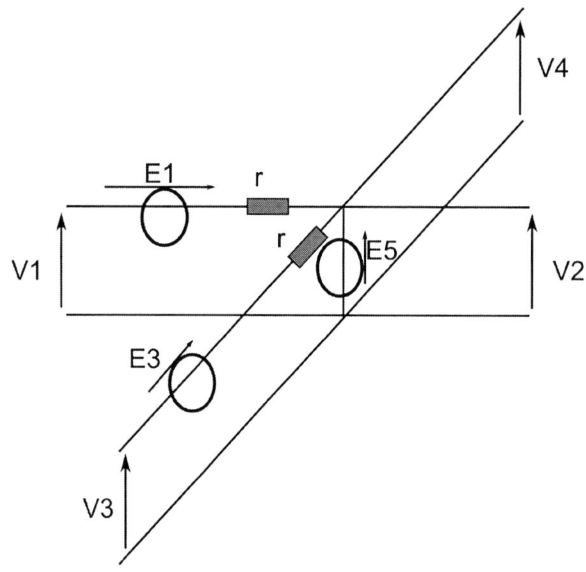

Figure 1.13. *Elementary cell with three sources E_1, E_3 and E_5*

Sources E_1, E_3 and E_5 rely upon electric walls so as to have an alternative current. If they rely upon magnetic and periodic walls the current will have a zero value.

We have eight equations and eight unknowns:

$$V_2 = V_4 = E_5 \quad [1.43]$$

$$I_1 + I_2 + I_3 + I_4 + I_5 = 0 \quad [1.44]$$

$$V_1 + E_1 - rI_1 - E_5 = 0 \quad [1.45]$$

$$V_3 + E_3 - rI_3 - E_5 = 0 \quad [1.46]$$

With the four equations for periodicity:

$$V_2 = V_1 e^{j\alpha} \quad [1.47]$$

$$I_2 = -I_1 e^{j\alpha} \quad [1.48]$$

$$V_4 = V_3 e^{j\beta} \quad [1.49]$$

$$I_4 = -I_2 e^{j\beta} \quad [1.50]$$

Upon rewriting these equations by eliminating V_2, I_2, V_4, I_4:

$$\begin{aligned} &I_1(1-e^{j\alpha}) + I_3(1-e^{j\beta}) + I_5 = 0 \\ &V_1 e^{j\alpha} = V_3 e^{j\beta} = E_5 \end{aligned} \quad [1.51]$$

$$\begin{aligned} &V_1(1-e^{j\alpha}) + E_1 - rI_1 = 0 \\ &V_3(1-e^{j\alpha}) + E_3 - rI_3 = 0 \end{aligned} \quad [1.52]$$

these two equations are written as:

$$\begin{aligned} &E_5 e^{-j\alpha}(1-e^{j\alpha}) + E_1 = rI_1 \\ &E_5 e^{-j\beta}(1-e^{j\beta}) + E_3 = rI_3 \\ &-rI_1(1-e^{j\alpha}) - rI_3(1-e^{j\beta}) = rI_5 \end{aligned} \quad [1.53]$$

That is to say:

$$\begin{aligned} &a = e^{-j\alpha}(1-e^{j\alpha}) \\ &b = e^{-j\beta}(1-e^{j\beta}) \end{aligned}$$

These equations then become:

$$r\begin{vmatrix} I_1 \\ I_3 \\ I_5 \end{vmatrix} = \begin{vmatrix} 1 & 0 & a \\ 0 & 1 & b \\ a^* & b^* & |a|^2+|b|^2 \end{vmatrix} \begin{vmatrix} E_1 \\ E_3 \\ E_5 \end{vmatrix} \qquad [1.54]$$

It is now necessary to move on to the calculation of actual values so as to determine the reflection coefficient.

$$\text{Hence: } \Gamma = 1 - \frac{2\frac{Z_0}{r}}{1+\frac{Z_0}{r}} YY^+ - \frac{2\frac{Z_0}{r}(1+|a|^2+|b|^2)}{1+\frac{Z_0}{r}(1+|a|^2+|b|^2)} ZZ^+ \qquad [1.55]$$

When r tends towards 0:

$$\Gamma = 1 - 2YY^+ - 2ZZ^+ \qquad [1.56]$$

For the matrix S:

$$S = \begin{cases} -1 \; short \; circuit \\ 1 \; open \; circuit \end{cases}$$

Once the Γ and S parameters have been determined, the iterative process may be applied with the aid of these equations.

$$\begin{aligned} B &= \Gamma A \\ A &= SB + A_0 \end{aligned} \qquad [1.57]$$

Upon achieving convergence, the field Ez may be determined across the entire structure.

To display the couplings between the cells, the transition from configuration to an activated line with a five-cell configuration, which is activated in Z (see Figure 1.14), is essential:

$$f = 3.10^9 GHz, Z_0 = 50\Omega, r = 5\Omega, \Delta l = \frac{\lambda}{10}$$

1.15. Validation examples

The structure below (Figure 1.14), is made up of 30*30 cells with a supply from the central cell (15:15) in *Y*.

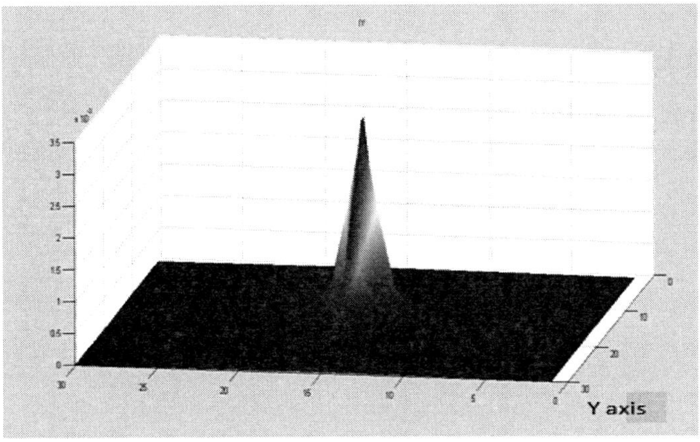

Figure 1.14. *Current density for the central cell activated in Y*

In Figure 1.15, the curve represents an emergence of the current from the central cell. The current density is of the order of 10^{-3} A/m². Across the rest of the cells the current is not, in fact, zero, but rather of the order of 10^{-5} A/m².

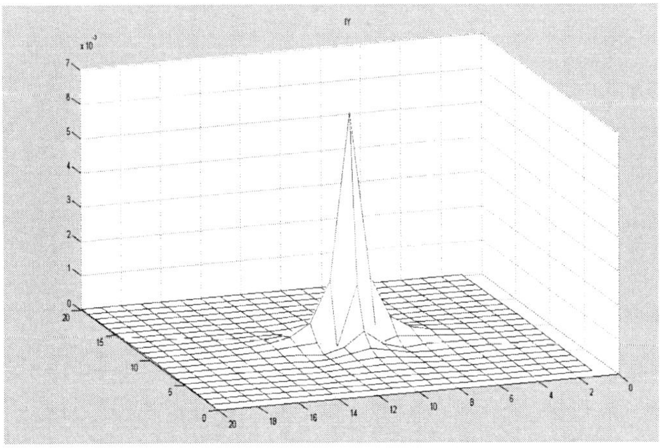

Figure 1.15. *Current density for a structure made up of 20 *20 cells with a central cell activated in Y*

This example in (Figure 1.15) demonstrates polarization in Y, with a source supplied in Y and short-circuited in X. The structure is 20*20 cells. The supply source is centered within a (10, 10) configuration with a current of the order of 6.65×10^{-5} A/m². A current emerges in X and Y (in Figure 1.16), since there is a coupling between X and Y.

To visualize couplings between cells, we transition from a single activated line circuit to a five-cell configuration, which is activated in Z (Figure 1.17).

In this configuration, for a given column of cells, these various cells are independent (in the Y direction), the equation being:

$$Z_L = \frac{Z_l - Z_0}{Z_l + Z_0}$$

with $Z_l = jlw$.

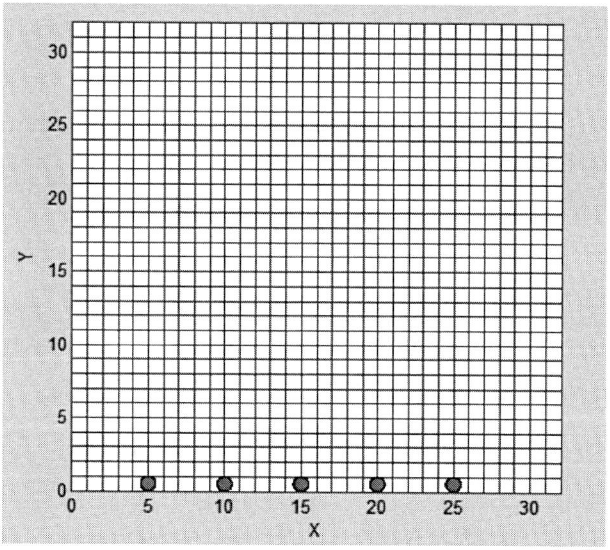

Figure 1.16. *Five-cell configuration with a source activated in Z*

The results are shown in (Figures 1.17 and 1.18). The field Ez is propagated according to the direction of polarization (*Y*) for each cell source which is fed in *Z*.

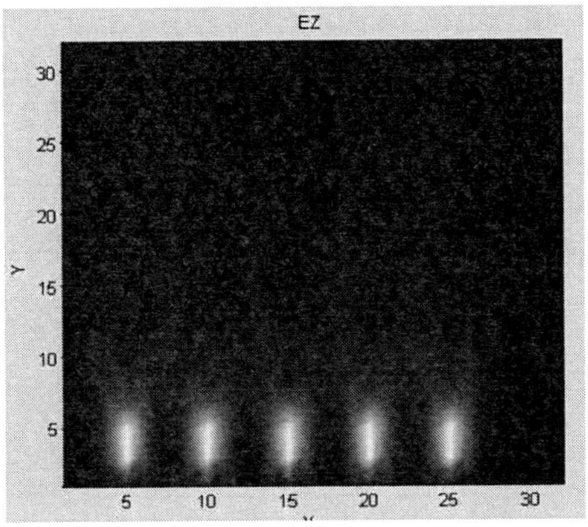

Figure 1.17. *Five-cell configuration with a source activated in Z, shown in two-dimensions*

In view of the presence of a resistance behind the auxiliary sources, the curve represents the decrease both in the magnitude of the field Ez (Figure 1.18) and the propagation following the *Y* direction.

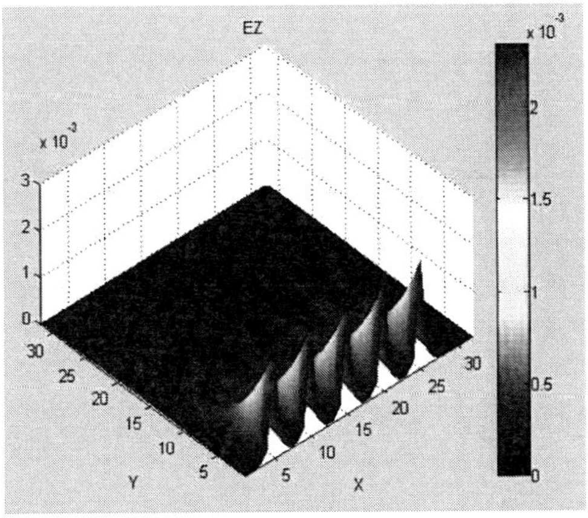

Figure 1.18. *Five-cell configuration with an activated source in Z shown in three-dimensions. For a color version of this figure, see www.iste.co.uk/baudrand/waves.zip*

That is to say, this is a bi-dimensional structure made up of 30 *30 cells, and the sources in Z on the first line are all activated. On the same line, the phase difference from one cell to another (depending upon the X direction) is of the order of $2\pi m/N$, with N being the number of cells on each line.

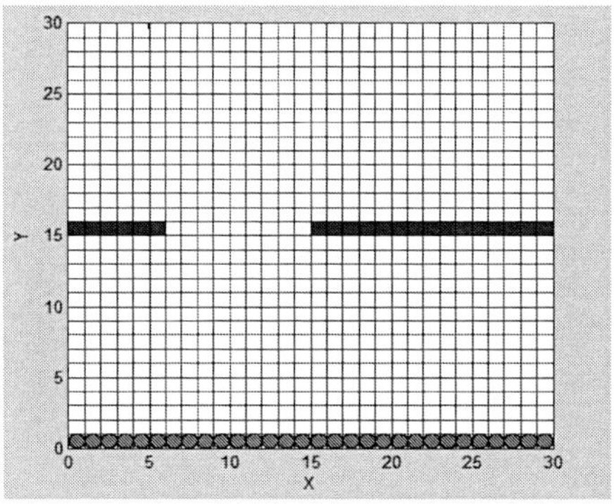

Figure 1.19. *Diagrammatic representation of the structure being studied. The source cells in red (lower line) are active and those in blue (above) are impediment cells. For a color version of this figure, see www.iste.co.uk/baudrand/waves.zip*

In this structure "impediment cells" will be inserted. These are cells covered with metal, at the 15th line, as shown in Figure 1.19. To simulate an "impediment cell", E_1 and E_3 are each replaced by an inductance and E_5 is replaced by a capacitance.

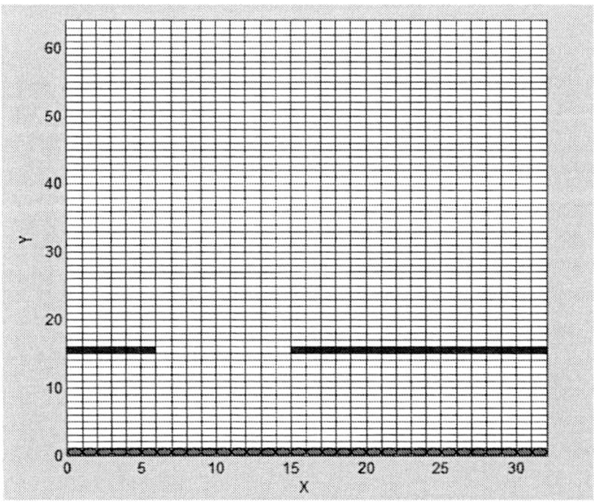

Figure 1.20. *Diagrammatic representation of the studied structure. The active source cells are in red (lower line) and the impediment cells are in blue (above). For a color version of this figure, see www.iste.co.uk/baudrand/waves.zip*

That is to say, this provides a configuration made up of 64*32 cells, each cell being of the size $dl = \lambda/20$. Our entire study revolves around a single elementary cell. The first-line sources of this structure are fed in Z. In X and Y, the sources are replaced by inductances known as L.

The impediment cells within the structure are on the 15th line. A slit was created on this line to examine the behavior of the Ez field (Figure 1.20).

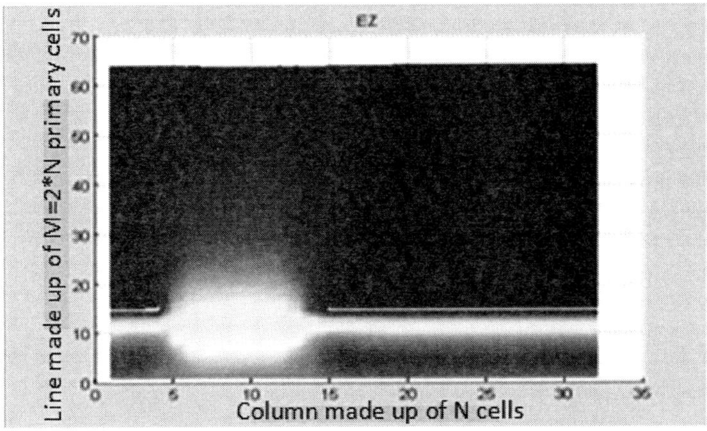

Figure 1.21. *E_z as a function of the number of cells in X and Y directions for a structure made up of 64x32 cells*

The curve shows the behavior of the density in the field Ez upon each cell, at the level of the impediments on line 15. The Ez fields are reduced while at the level of the slit, the field is propagated and this shows stationary waves.

Within the following curve (Figure 1.22), the behavior of the Ez field at the level of the 15th line is shown: being metal from 0 to 5, followed by an opening from 5 to 15, and then metal from 15 to 32.

Let us now take the same configuration previously studied and invert the cells, by replacing the cells at the slit by impediment cells (Figure 1.23). The propagation of the Ez field is shown in Figures 1.24 and 1.25.

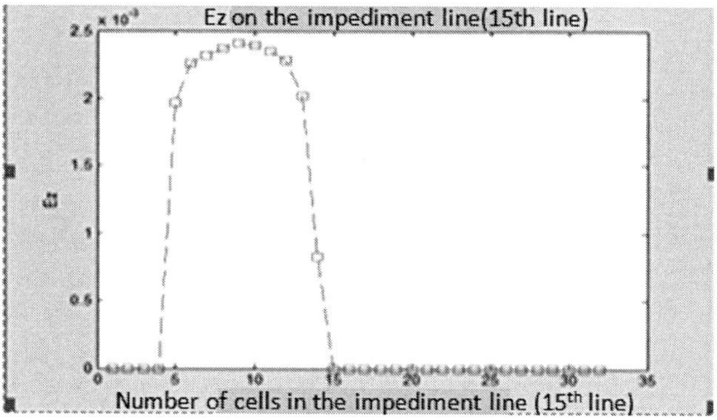

Figure 1.22. *Behavior of the E_z field on the 15th line*

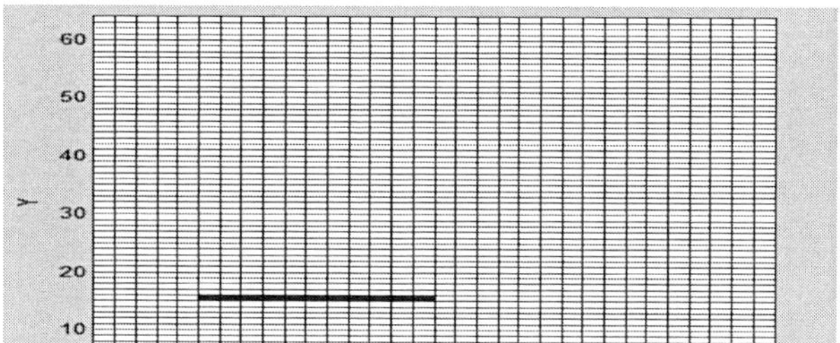

Figure 1.23. *Diagrammatic representation of the structure studied. The active source cells are shown in red (lower line) and the impediment cells are shown in blue (above). For a color version of this figure, see www.iste.co.uk/baudrand/waves.zip*

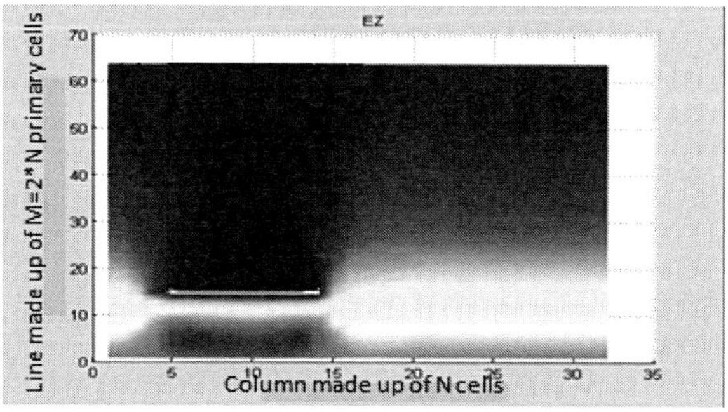

Figure 1.24. *E_z as a function of the cells number in X and Y directions for structure made up of 64 x 32 cells and shown in Figure 23. For a color version of this figure, see www.iste.co.uk/baudrand/waves.zip*

The behavior of the field Ez differs from that found within the IRIS structure; at the level of impediments on line 15 the field Ez backscatters towards the first line, but at the level of the slit for the stationary waves all along the remainder of the structure. The behavior of the Ez fields at the level of the 15th line is as follows: a gap from 0 to 5, then metal from 5 to 15, and then a gap from 15 to 32. This is shown on Figure 1.25.

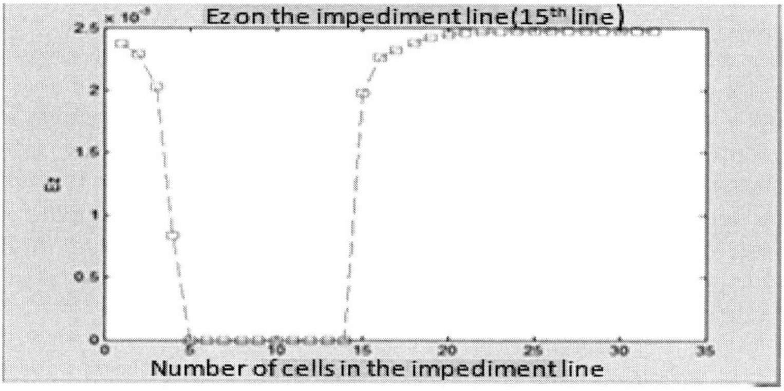

Figure 1.25. *Behavior of the E_z field on the 15th line*

The case of the central cell, the source being fed in Z

This sub-heading refers to the case of a central cell fed by a source in *Z* (Figure 1.26). Here, the sources within *X* and *Y* are replaced by inductances. We see the behavior of field Ez across a structure of 64 *64 cells with a cell length of the order of $\lambda/20$.

When the source in *Z* of the center cell is activated, the field Ez is propagated in the two directions *X* and *Y* by creating both circular and stationary waves around this same cell (Figure 1.27). The distance between each wave is of the order of 20 cells which makes it possible to find the start wavelength λ, since $dl = \lambda/20$.

Once the central cell source in *Z* has been activated, a symmetrical line around the propagation of Ez appears (Figure 1.28) following the *x* direction (the curve in red, boxes) and another symmetrical line around the propagation of Ez following *y* (the curve in blue, crosses) will be shown.

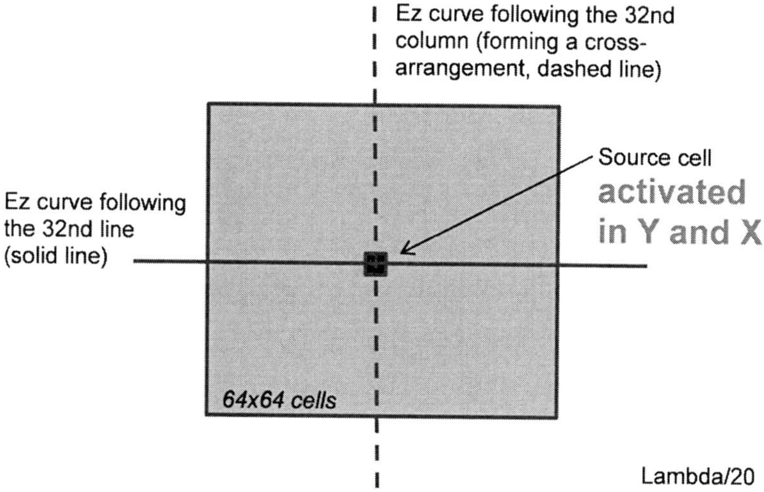

Figure 1.26. *Description of the configuration 64 *64 cells*

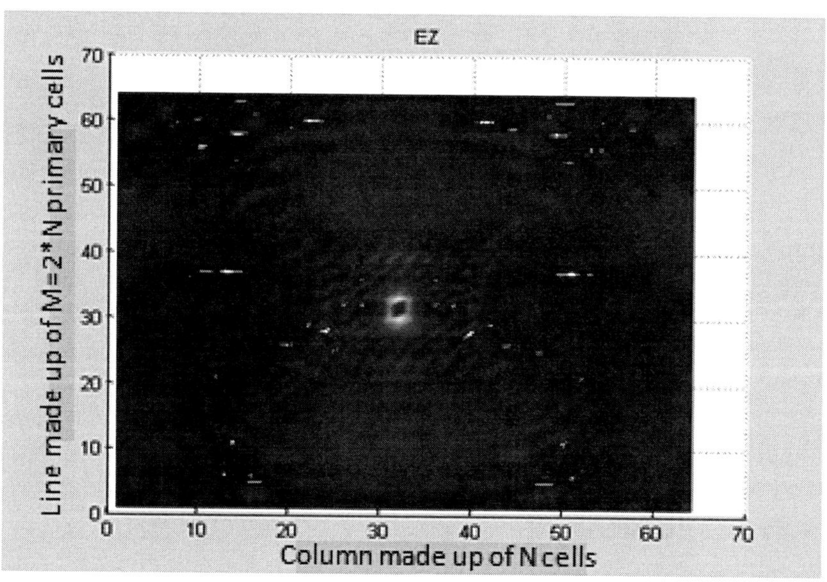

Figure 1.27. *Field E_z as a function of the cells number in X and Y directions for a structure made up of 64*64 cells and shown in Figure 1.26 with dl= $\lambda/20$*

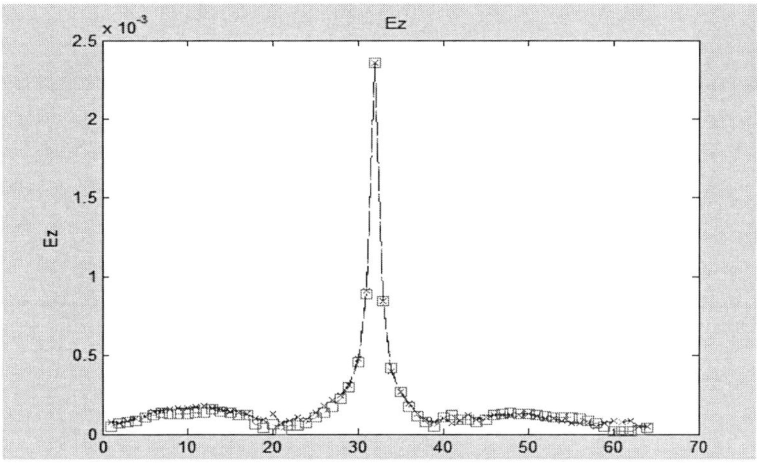

Figure 1.28. *Behavior of field Ez at the level of the central source in X (red boxes) and in Y (blue crosses). For a color version of this figure, see www.iste.co.uk/baudrand/waves.zip*

For a length $dl = \lambda/100$; this is the source that is fed in Z – the central cell. The curve shows the appearance of the field Ez (see Figure 1.29) which is propagated from the central source across the entire structure of 64 *64 cells, with a cell length of the order of $\lambda/100$.

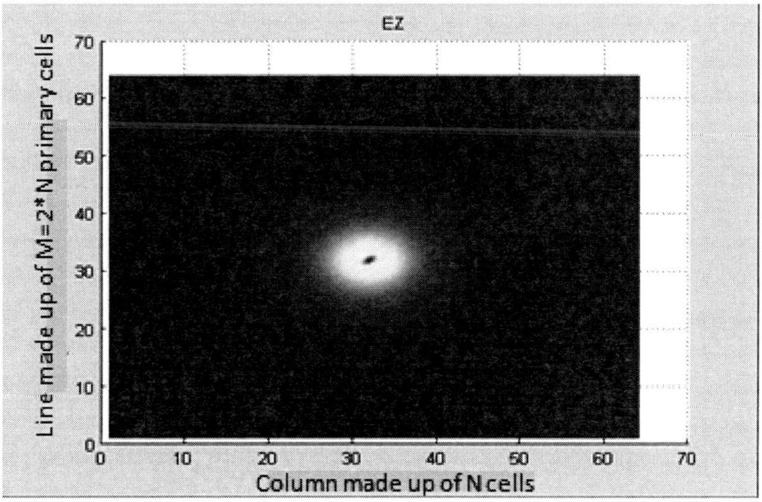

Figure 1.29. *Field E_z as a function of the number of cells in X and Y directions for a structure made up of 64*64 cells and shown in Figure 1.26 with dl= λ/100*

The validation of the study of this type of bi-dimensional configuration must undergo a comparison with studies previously completed, such as the Hankel function (Figure 1.30). This is used in applications for electromagnetic cylindrical guides, for example, within a given diffraction problem, by a cylinder which is both infinite in length and illuminated by a plane wave. The Hankel function tends towards 0 by moving away from the center.

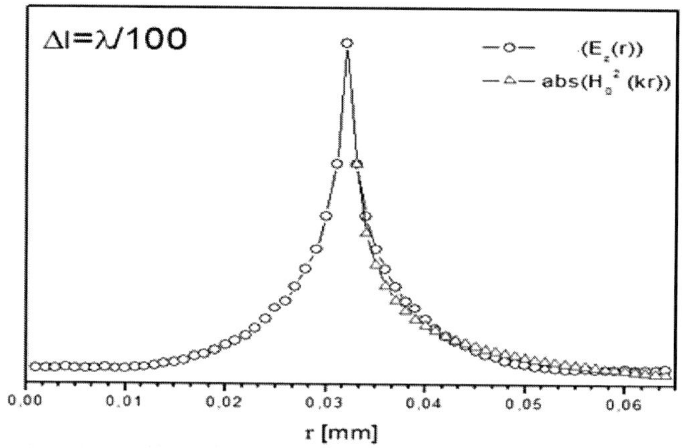

Figure 1.30. *Comparison of the Ez field's behavior for a central source which is active in Z, and the Hankel function*

The Ez field is at its maximum at the level of the central source, and decreases by being propagated along the structure in both directions *X* and *Y*. It is superimposed with the Hankel function for a cell length of the order of $\lambda/100$. This both verifies and validates the theoretical approach adopted for the study of quasi-periodic bi-dimensional structures [AZI 13].

1.16. Lenses and meta-materials

Isolating the elementary cell of the bi-dimensional structure (Figure 1.31) is necessary to calculate the values of permittivity and the corresponding permeability. The entire study of the structure takes place across the elementary cell, viewing the periodicity of the walls surrounding the latter.

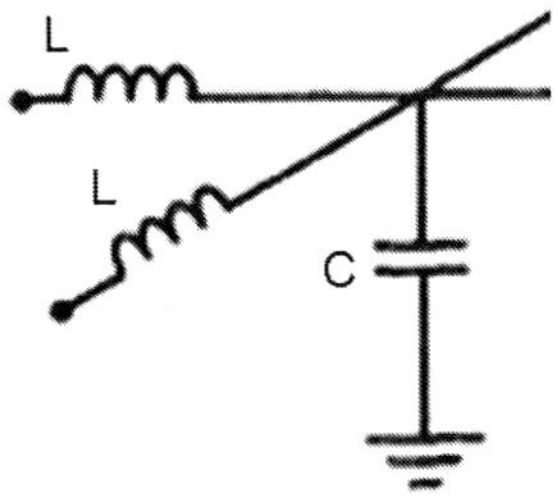

Figure 1.31. *Elementary cell*

The values of L and C determine the values of μ and ε, that is to say δl the length of the elementary cell, the capacity per unit of length is equal to:

$$C_c = C/\delta l \quad [1.58a]$$

$$L_L = L/\delta l \quad [1.58b]$$

and:

$$\sqrt{L_L C_C} = \sqrt{\mu_0 \varepsilon_0 \varepsilon_r} \quad [1.58c]$$

Assuming $\mu = \mu_0$ the refraction index is given:

$$n = \sqrt{\varepsilon_r} = \sqrt{\frac{L_L C_C}{\mu_0 \varepsilon_0}} = \frac{1}{\delta l}\sqrt{\frac{LC}{\mu_0 \varepsilon_0}} = \frac{c}{\delta l}\sqrt{LC} \quad [1.59a]$$

Where c = 3.10^8 m/s is the speed of light.

C_0 and L_0 are taken to be the values which lead to the formula n=1

$$\sqrt{\frac{L_0}{C_0}} = \sqrt{\frac{\mu_0}{\varepsilon_0}} \quad [1.59b]$$

With

$$\sqrt{L_0C_0} = \sqrt{\mu_0\varepsilon_0}\delta l \qquad [1.59c]$$

Equation [1.59c] is extracted from the formula in [1.58c]

Multiplying [1.59b] by [1.59c] results in obtaining L_0 as:

$$L_0 = \sqrt{\frac{\mu_0}{\varepsilon_0}}\sqrt{\mu_0\varepsilon_0}\delta l = \mu_0\delta l \qquad [1.60a]$$

Likewise, dividing [1.59c] by [1.59b] results in obtaining C_0 as:

$$C_0 = \varepsilon_0\delta l \qquad [1.60b]$$

On the other hand, C will evolve. With n as an index C becomes

$$\mathrm{C} = C_0 n^2 \qquad [1.61]$$

Conventional electrodynamics impose a resolution limit when conventional imaging uses lenses. This fundamental limit, called the "diffraction limit", in its ultimate form, is attributed to the wavelength of electromagnetic waves. The loss of resolution, which is a valid hypothesis, even if the diameter of the lens proves to be infinite, constitutes the source of the diffraction limit in its ultimate form. In the case of projecting an image from a punctual source, the diffraction limit appears as an unclear image upon a surface area of approximately a wavelength of the diameter of the lens:

$$\Delta\rho \sim 2\pi / k_0 = \lambda \qquad [1.62]$$

In 2000, John Pendry elaborated a new analysis of Veselago's lens (Figure 1.32) by introducing evanescent waves and observed that these lenses might indeed overcome the diffraction limit [PEN 99].

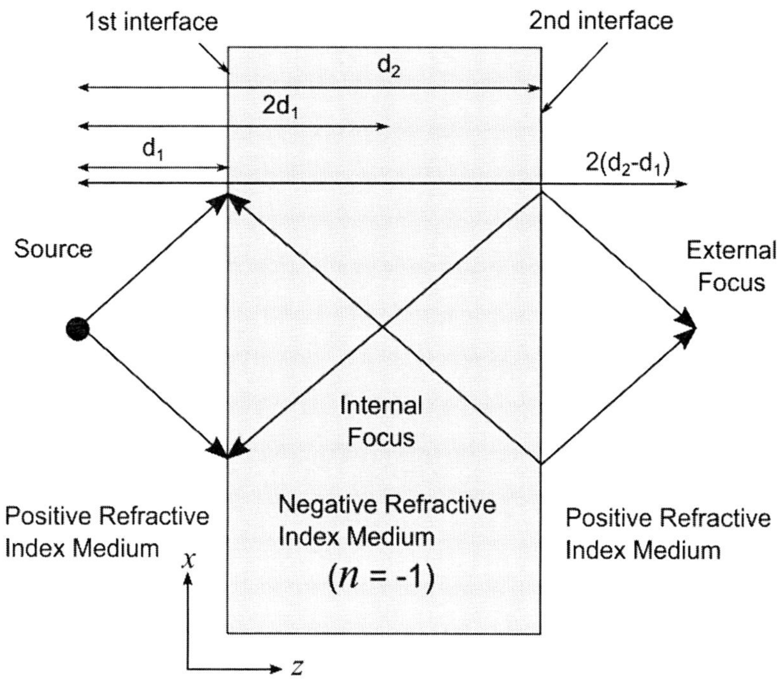

Figure 1.32. *Meta-material lens*

Pendry suggested that Veselago's lens makes it possible to have a perfect image, if it is entirely without any loss and its refractive index is exactly equal to −1 in relation to the surrounding environment. The left hand lens realizes a super-resolution image by placing the emphasis upon waves, which are propagated in the way that using a conventional lens would allow. However, it furthermore takes the amplification and restoration of evanescent waves which are splitting up and coming from the source. This restoration of evanescent waves, within the image plane extends to the numbers of maximal accessible waves, making it possible to create a very-high resolution image. The physical mechanism behind the growth of evanescent waves is somewhat interesting. Within Negative Refractive

Index (NRI) – (left-handed) materials and multiple reflections lead both to the increase and decrease of evanescent waves. Nevertheless, the index of $n = -1$ corresponds to a resonance phenomenon in which the reduction solution is canceled out, thus simply leaving the current growth of these waves.

This is achieved because when $n = -1$, the second NRI/PRI interface (Positive Refractive Index (PRI)) within Figure 1.1 corresponds to an infinite reflectance, while the first PRI/NRI interface is adjusted. In one sense, one may assimilate Veselago's lens as an inverse system which precisely restores wave propagation within free space.

An image of the plane version of Veselago's lens which was built at the University of Toronto is shown in Figure 1.33 [IYE 02]. The NRI lens is a grill structure made up of 5 × 19 cells of printed micro-strip bands, operated with series C_0 capacitors and short-circuited with an L_0 inductance. This NRI structure is sandwiched between two printed discharged grids which act as a homogeneous environment having a positive refraction index. The first non-charged network is excited using a monopole (a punctual source) which is fixed to the left-hand grid. This is reproduced by the NRI lens on the second grid. The vertical electric field across the entire structure is measured with the aid of a detection probe.

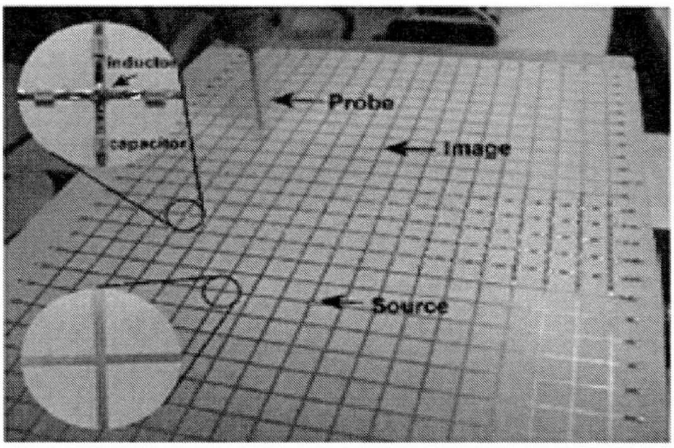

Figure 1.33. *Structure of the Veselago lens*

Meta-material cells are made up of capacitances and of self-induction coils, as shown in Figure 1.34.

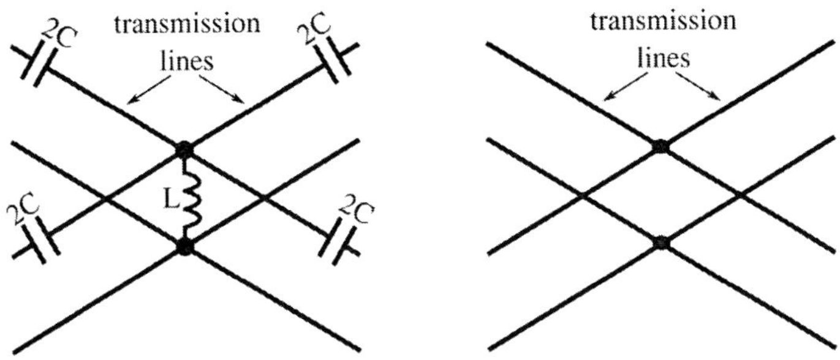

Figure 1.34. *Transmission line elementary cells*

To simulate the configuration (Figure 1.35), a source cell will be placed level with the third line at a distance from the primary interface of the meta-materials of the order *dl* = 2**d* (that is to say corresponding with two lines of cells). The second interface [PEN 00] is placed at a distance of 2**d* from the first interface. The source image should be focused on a distance of 2**d* behind the second interface to validate this study [IYE 03].

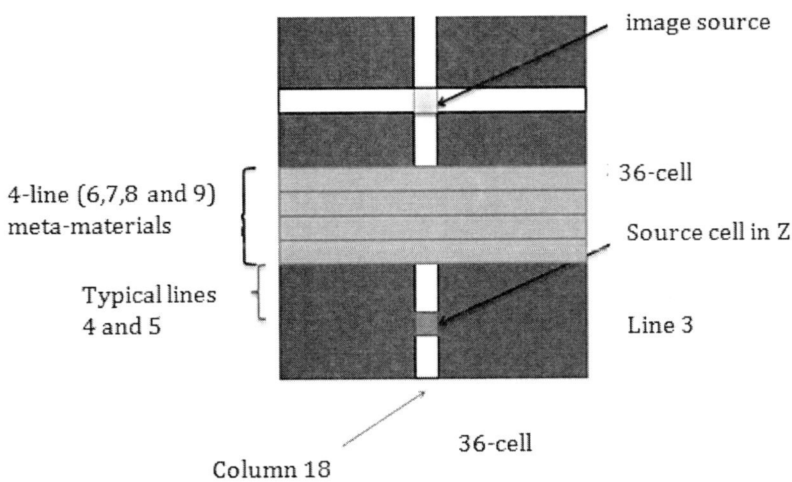

Figure 1.35. *Configuration of the lens*

Figure 1.36 shows the behavior of evanescent waves which increase within the lens. The dashed lines indicate the location of the NRI region and the solid lines indicate the location of the source (on the top) and the external image is also given (see the view on the bottom).

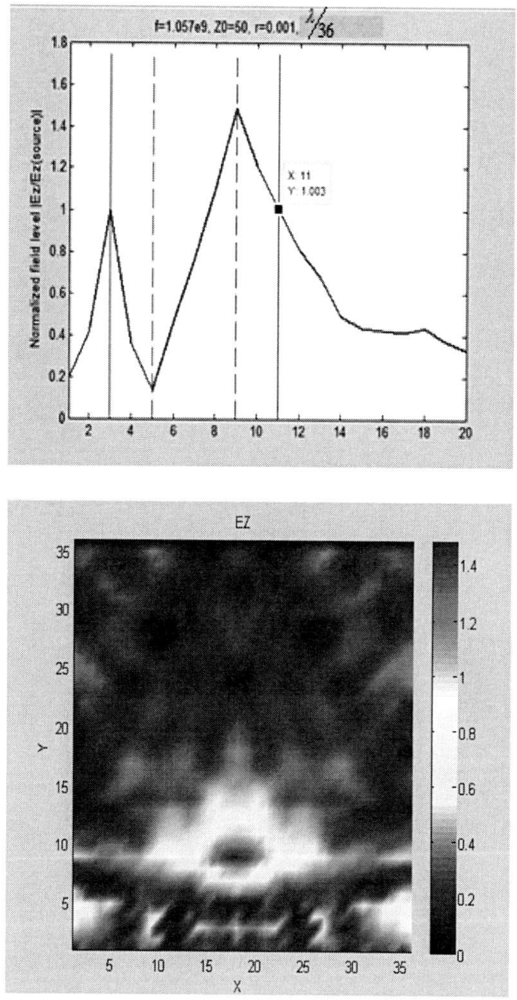

Figure 1.36. *Behavior of the field |Ez|. For a color version of this figure, see www.iste.co.uk/baudrand/waves.zip*

This thus represents an optical system which is likely to achieve a perfect image [IYE 03] of an unlimited object caused by diffraction. Negative refraction ends up forming perfect images, provided that the period of the material is sufficiently weak when compared to wavelength. Figure 1.36 shows the principle of the super lens [GRB 04].

The elimination of the limit due to diffraction results from the amplification of evanescent waves. This amplification, which is caused by the inversion of the wave vector, makes it possible for the evanescent wave to be transmitted to the image without any reduction. To obtain this result, it is necessary for the transmission to take place without any reflection occurring upon the interfaces with the external conventional environment. If the latter consists of either air or a vacuum, with the formula $\varepsilon r = \mu r = 1$, it is therefore necessary that the left-hand material has the formula $\varepsilon r = -1$ and $\mu r = -1$.

1.17. Conclusion

We have, throughout this work, been able to trace the development of design codes and appropriate methods to deal entirely with different types of quasi-periodic structures. At the start, our objective was to both improve and broaden the scope of the WCIP iterative method within both the one-dimensional and bi-dimensional sphere. The developments introduced in the iterative method made it possible to attribute to it some qualities which are necessary to resolve propagation issues within any environment (whether homogeneous or inhomogeneous), as well as within quasi-periodic complex structures.

A study of quasi-periodic structures was made by determining the behavior of current density across each cell. In this section, we have limited the discussion to one-dimensional quasi-periodic structures. Examination of the general application of the concept of the formalism of quasi-periodic structures and the study of bi-dimensional structures was completed. When dealing with magnetic field behavior, a validation was carried out of a bi-dimensional design made up of an artificial line fitted with sources which were activated upon each cell of the structure. The last section was devoted to the study of the applications of bi-dimensional quasi-periodic configurations. Indeed, the use of the iterative method, with its novel approach with hexapole structures, has allowed us to study the so-called IRIS system as a primary application. Owing to the analogy with photonic systems with infinite cylinders, a comparison with Hankel's function has made it possible to validate all of our work. One final application which advances the provision of the iterative method is that of the meta-material flat lens.

This work opens up a multitude of pathways. For example, the study of quasi-periodic structures based upon new materials, such as graphene, by using the WCIP method in this field appears somewhat interesting. Indeed, owing both to the exceptional properties that these materials offer and to the WCIP method, we will now be able to study the behavior of some complex structures, such as meta-surface structures, inhomogeneous structures and meta-material structures.

2

Formulation and Validation of the WCIP Applied to the Analysis of Multilayer Planar Circuits

2.1. Introduction

For those working within the electromagnetic sphere, wave propagation in multilayer structures is a subject which has proven to be both somewhat significant and of great interest [SUT 03]. RF circuits and microwaves are generally produced within a single layer configuration, although multilayer circuit structures have mainly been used for low-frequency digital systems. The use of multilayer planar circuits makes it possible both to produce more compact microwaves and more flexibly-designed circuits [MIL 07]. The wave concept is a well-established procedure for dealing with electromagnetic problems [KUR 65]. It is used in the Transmission Line Matrix (TLM) method [FIC 09]. To reduce both time limits and development costs of monolithic hyper-frequency integrated circuits Monolithic Microwave Integrated Circuits (MMIC), it is of paramount importance to use a rapid and efficient tool, which may accurately predict both the electrical and electromagnetic behavior of a given device. This method is not restricted by circuit complexity and has proven to be particularly interesting in the case of planar circuits. The Wave Concept Iterative Procedure (WCIP) approach consists of separating the structure

Chapter written by Alexandre Jean René SERRES and Georgina Karla DE FREITAS SERRES.

studied by an environment which is both superior and inferior, although both acting homogeneously across an interface, in the case of a planar circuit with a single interface. Figure 2.1 shows an example of a multilayer planar circuit with plating on several interfaces.

The boundary conditions operating across interfaces are shown by a diffraction operator, $S_{\Omega i}$, which is defined within the spatial sphere. Within homogeneous environments, wave propagation between interfaces is shown by a transfer matrix *T*, and across the extremes of the circuit (both at the highest and the lowest parts) by a reflection operator, Γi. This propagation is defined as being within the modal sphere. The method supplies a mixed resolution within the modal and spatial sphere and takes full advantage of each sphere with a low computation time.

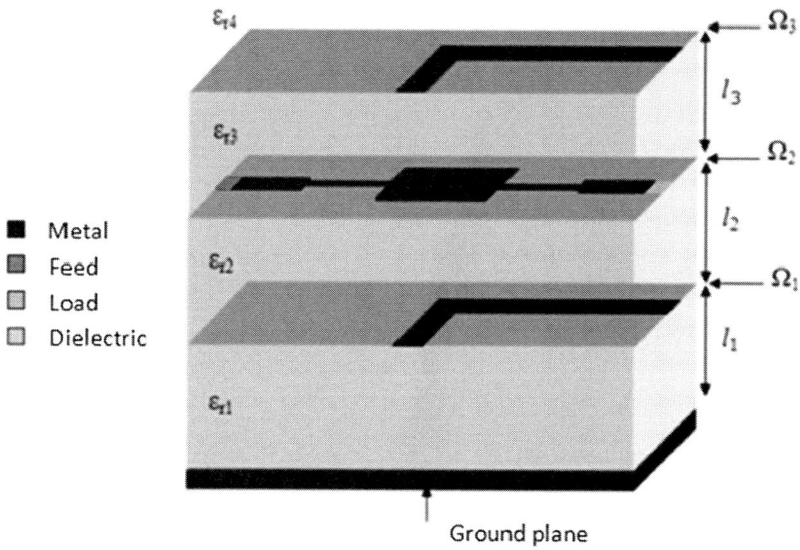

Figure 2.1. *Example of the multilayer planar structure. For a color version of this figure, see www.iste.co.uk/baudrand/waves.zip*

In this Chapter, section 2.2 provides theoretical formulation with simulation results from two case studies. An original application of the concept is shown in section 2.3, together with the theoretical formulations and simulations for an ideal polarizer using the WCIP. In section 2.4, a multilayer structure as a quasi-optic compact amplifier is suggested with a detailed analysis.

2.2. WCIP formulation

2.2.1. *Multilayer formulation*

The method described here is based upon the complete transverse wave formulation in which the concepts of dual quantities, current density and electric field, are considered. It was brought in as a means to express the boundary conditions across the air/dielectric interface in terms of waves. The waves within both the spatial and modal spheres are directly inferred by a Fourier Modal Transform (FMT) and its converse transform the so-called Inverse Fourier Modal Transform (or FMT^{-1}). We will now consider the general issue with a single interface, as shown in Figure 2.2 by means of its cross-section.

The source A_{0i} generates waves, one on each side of the relevant interface Ω_i.

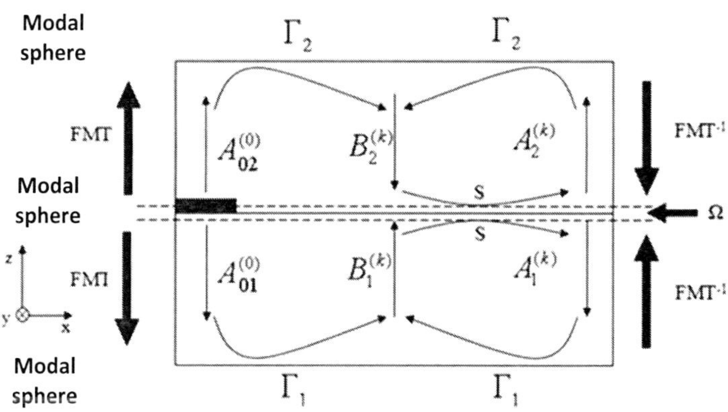

Figure 2.2. *Behavior of waves A and B, the transformation spheres within the iterative process for a single interface*

The incident wave (A) and the reflective wave (B), the electric field (E) and the current density (J) may be linked by:

$$\vec{A}_i = \frac{1}{2\sqrt{Z_{0i}}}\left[\vec{E}_i + Z_{0i}\vec{J}_i\right] \qquad [2.1]$$

$$\vec{B}_i = \frac{1}{2\sqrt{Z_{0i}}}\left[\vec{E} - Z_{0i}\vec{J}_i\right] \qquad [2.2]$$

where i indicates the environment 1 or 2 corresponding to a given interface in Ω (resistance). Z_{0i} is the impedance which is characteristic of the same environment i with $\vec{J}_i$ being the surface current density vector.

The waves within both the spatial and modal spheres are directly inferred by the FMT and its converse transform, the so-called FMT^{-1}. The decomposition of the electromagnetic wave in propagated modes (both evanescent and propagating) within a waveguide containing periodic walls (modes TE and TM) occur through the application of the FMT. The FMT is made up of the algorithm 2D-FFT and the transformation from the spectral sphere to the modal sphere. The conversion of the so-called modal Fourier and its converse ensures the transition between the two spheres.

The diagram normally used for the WCIP, in respect to a single layer, is very simple. Two operators relate both incident and reflective waves in the spatial sphere and the spectral sphere governs the iterative procedure. It may be shown by two equations.

$$\vec{A} = S\vec{B} + A_0 \qquad [2.3]$$

$$\vec{B} = \Gamma\vec{A} \qquad [2.4]$$

When the wave concept is extended to multilayered structures, as Figure 2.3 shows. A new formulation between waves ($A_{P\text{-}1}$, $B_{P\text{-}1}$) and (A_p, B_p) is introduced for the transition between two adjacent interfaces [SER 10].

This formulation is obtained from the TLM impedance. The transfer matrix between two adjacent interfaces is expressed as:

$$\begin{bmatrix} B_{1,p}^{\alpha}(k_x,k_y) \\ B_{2,p-1}^{\alpha}(k_x,k_y) \end{bmatrix} = \begin{bmatrix} T_{11}^{\alpha} & T_{12}^{\alpha} \\ T_{21}^{\alpha} & T_{22}^{\alpha} \end{bmatrix} \begin{bmatrix} A_{1,p}^{\alpha}(k_x,k_y) \\ A_{2,p}^{\alpha}(k_x,k_y) \end{bmatrix} \qquad [2.5]$$

or, alternatively:

$$[T^{(\alpha)}] = \begin{bmatrix} \dfrac{(Z_c^{(\alpha)2} - Z_{0i}^2)\sinh(\gamma l)}{2Z_c^{(\alpha)}Z_{0i}\cosh(\gamma l) + (Z_c^{(\alpha)2} + Z_{0i}^2)\sinh(\gamma l)} & \dfrac{2Z_c^{(\alpha)}Z_{0i}}{2Z_c^{(\alpha)}Z_{0i}\cosh(\gamma l) + (Z_c^{(\alpha)2} + Z_{0i}^2)\sinh(\gamma l)} \\ \dfrac{2Z_c^{(\alpha)}Z_{0i}}{2Z_c^{(\alpha)}Z_{0i}\cosh(\gamma l) + (Z_c^{(\alpha)2} + Z_{0i}^2)\sinh(\gamma l)} & \dfrac{(Z_c^{(\alpha)2} - Z_{0i}^2)\sinh(\gamma l)}{2Z_c^{(\alpha)}Z_{0i}\cosh(\gamma l) + (Z_c^{(\alpha)2} + Z_{0i}^2)\sinh(\gamma l)} \end{bmatrix} \qquad [2.6]$$

where Z_c is the mode impedance, Z_{0i} the environment impedance i, γ the transmission factor and l the height between two adjacent interfaces.

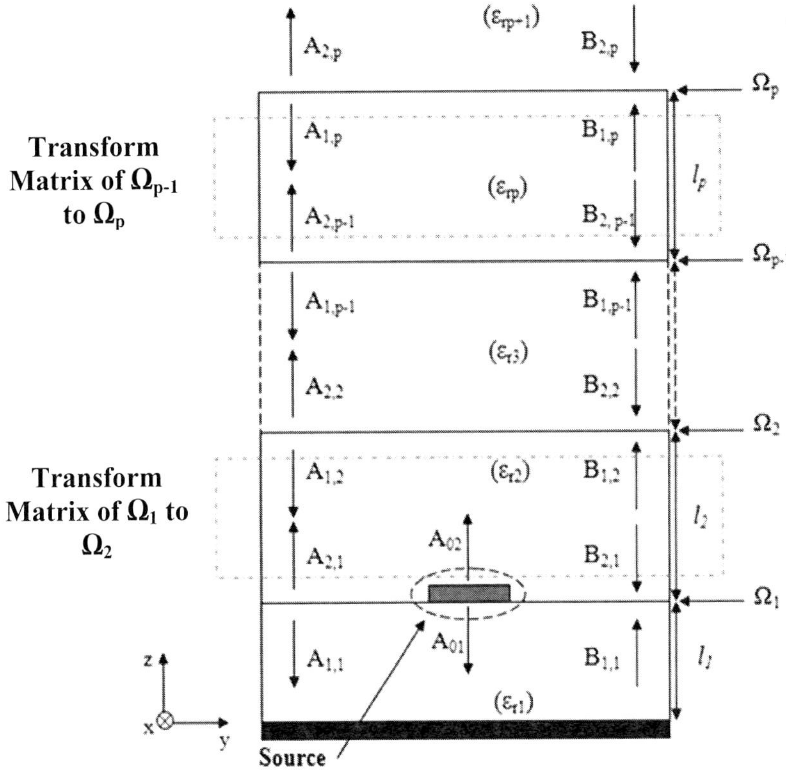

Figure 2.3. *Transfer matrix method, waves and the source within a multilayer structure*

Finally, it is possible to state boundary conditions in terms of waves across every cell, and the tangential electric and magnetic fields can then be calculated from the following equations:

$$\vec{E}_i = \sqrt{Z_{0i}}\left(\vec{A}_i + \vec{B}_i\right) \qquad [2.7]$$

$$\vec{J}_i = \frac{1}{\sqrt{Z_{0i}}}\left(\vec{A}_i\ \vec{B}_i\right) \qquad [2.8]$$

2.2.2. *Simulation results*

To show and evaluate the performance of the proposed tools, two case studies were completed. The cases considered made it possible to deal with both coupling problems and propagation issues.

2.2.2.1. *Structure with a gap and a superimposed resonator*

A simulation of a microstrip structure with a gap and a superimposed resonator was completed. Figure 2.4 shows the multilayer structure containing various interfaces.

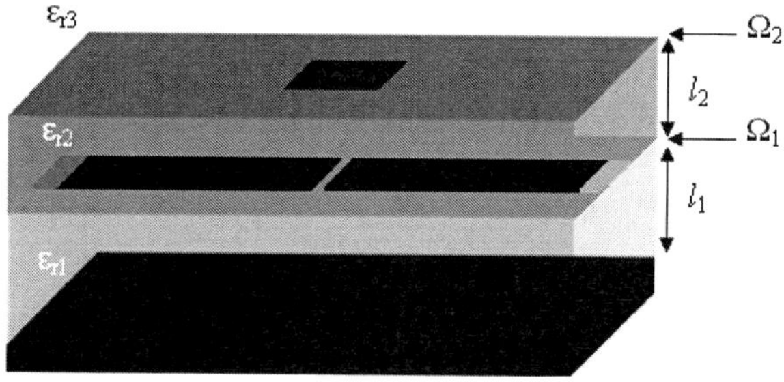

Figure 2.4. *3-D view of the multilayer structure*

On the first interface, the two microstrip lines have a width of 25.9 mm, a length of 2.3 mm with a gap of 1 mm. On the second interface, the resonator has a width of 2.3 mm and a length of 27.3 mm. The waveguide is 20 mm in width and 60 mm long, $\varepsilon_{r1} = \varepsilon_{r2} = 2.33$, $\varepsilon_{r3} = 1$, $L_2 = L_3 = 0.8382$ mm. Both the simulated source for an internal effective impedance of 50 Ω and the simulated charge of the 50 Ω have dimensions of 2.3 mm × 2.3 mm.

The current density *Jx* upon the resonator of 4 GHz is shown in Figure 2.5. The anticipated physical behavior is clearly evident.

As may be seen in Figure 2.5, the reflection coefficient $|S_{11}|$ and the transmission coefficient $|S_{21}|$ have values which are indeed very close to the measurement results. In particular, the resonance frequency obtained by the WCIP (at 4.05 GHz) is closer to the measurement (at 4.08 GHz) than the simulation (at 4.11 GHz in [FIL 00]), as much in terms of frequency as amplitude. The calculation time for a complete simulation is approximately 20 min with a 2.53 GHz Intel Core 2 Duo processor.

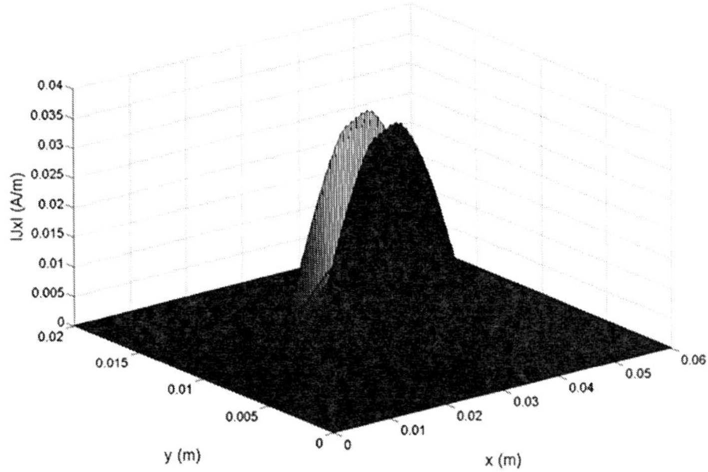

Figure 2.5. *Behavior of current density on the second interface at 4 GHz. For a color version of this figure, see www.iste.co.uk/baudrand/waves.zip*

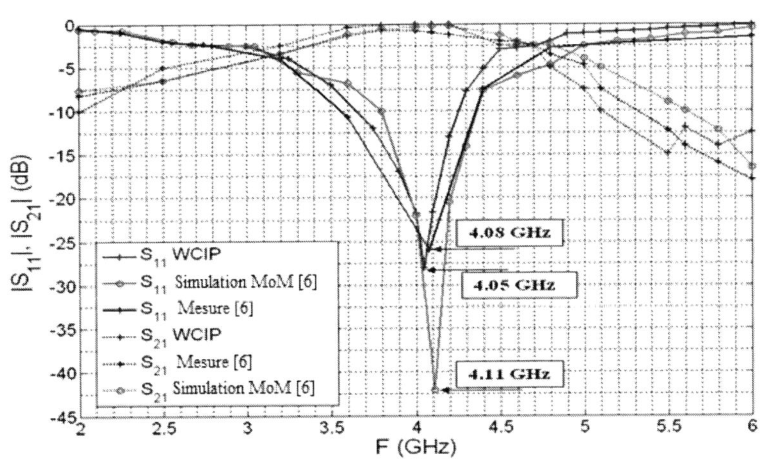

Figure 2.6. *Absolute value of the reflection coefficient, $|S_{11}|$ and transmission coefficient $|S_{21}|$ in dB according to the frequency in GHz. For a color version of this figure, see www.iste.co.uk/baudrand/waves.zip*

2.2.2.2. *Patch antenna with slit interfaces*

A patch antenna with a combined slit arrangement was simulated with the software program MatLab. Figures 2.7 and 2.8 show the multilayer structure with the various interfaces. On the first interface, the microstrip power line has a width of 1.9 mm and a length of 14.7 mm. On the second interface, the slit has a length of 4.7 mm and a length of 0.6 mm. The patch on the third interface has a width of 16 mm and a length of 6.65 mm. The wave guide is 22 mm in width and 22 mm length, $\varepsilon_{r1} = \varepsilon_{r4} = 1$, $\varepsilon_{r2} = \varepsilon_{r3} = 3.38$, $l_2 = l_3 = 0.81$ mm.

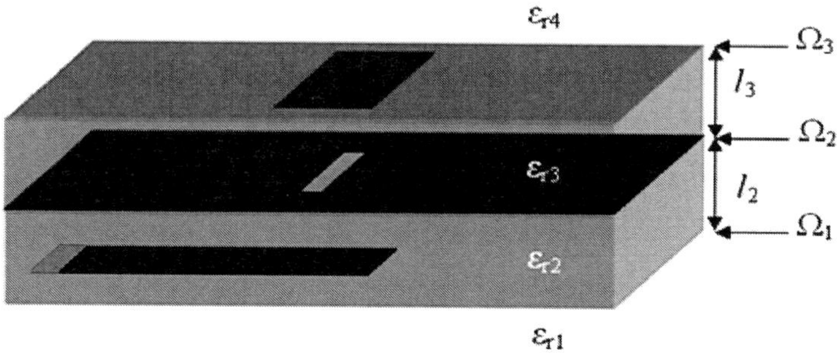

Figure 2.7. *3-D view of the patch antenna structure containing slit interfaces*

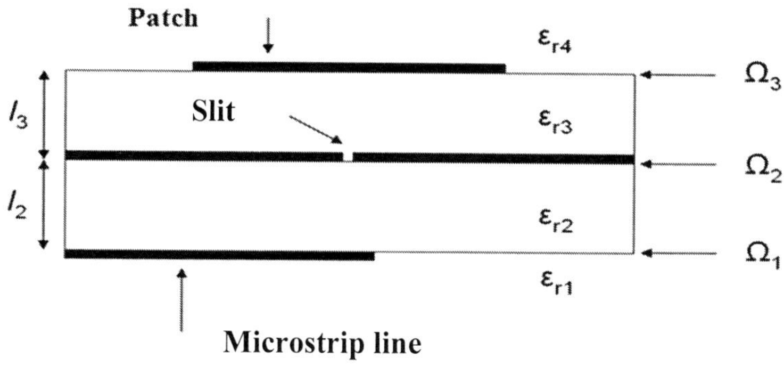

Figure 2.8. *Cross-section of the antenna structure*

Figures 2.9 and 2.10 show the behavior of current densities upon the second and third interface with a frequency of 10 GHz. The influence of the first interface, which contains the feed line, is seen in Figure 2.9. The expected physical behavior may be observed around the dielectric opening (or slit) with an increase in current density, followed by a reduction up to the surfaces of the waveguide.

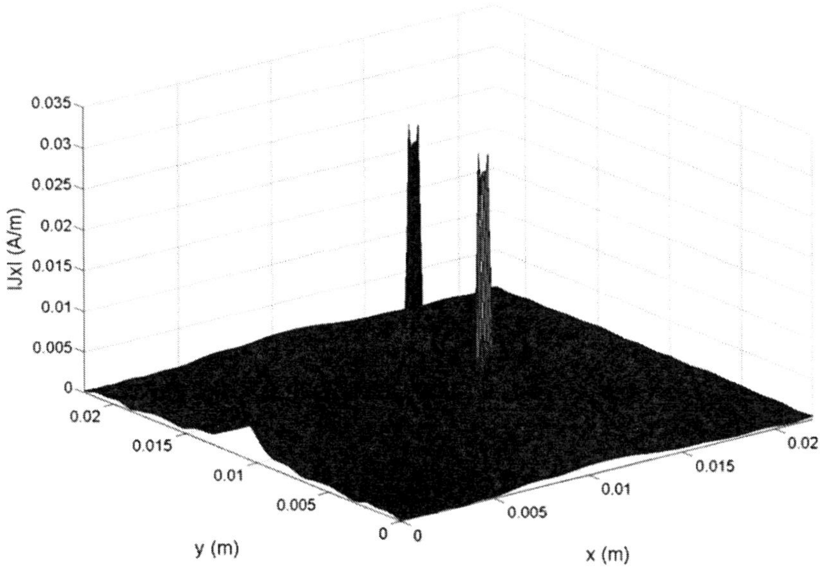

Figure 2.9. *Behavior of current density across the second 10 GHz interface. For a color version of this figure, see www.iste.co.uk/baudrand/waves.zip*

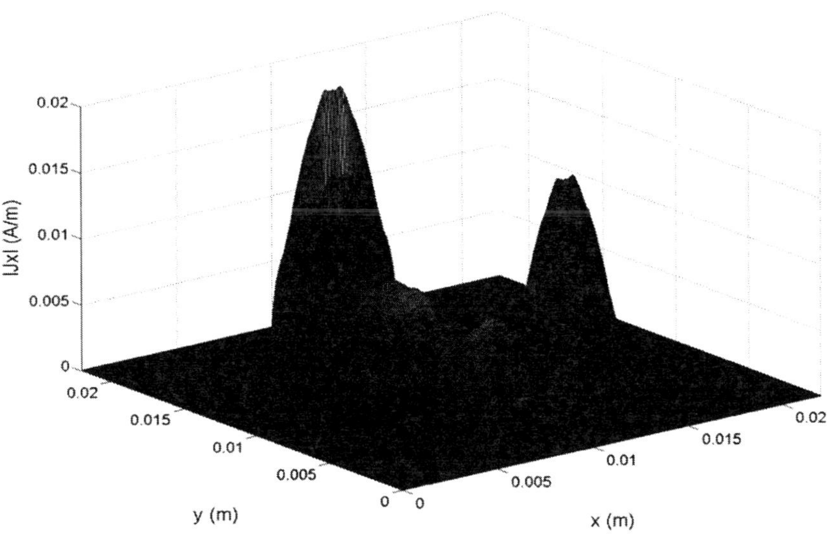

Figure 2.10. *Behavior of the current density upon the third interface at 10 GHz. For a color version of this figure, see www.iste.co.uk/baudrand/waves.zip*

The behavior of current density is shown in Figure 2.10, which shows that the frequency of the patch resonance is close to 10 GHz. Furthermore, the influence of the dielectric opening (or slit) in the second interface may be observed.

As can be seen in Figure 2.11 (the blue curve), the resonance frequencies of simulated results are very close to the measured frequencies. Moreover this agrees, in terms of amplitude, with the first resonance frequency which is shown in [ALI 98]. Consequently, it can be confirmed that the method suggested here gives satisfactory results, as such results are closer than the measurements of the simulations, when using the Finite-Difference-Time-Domain (FDTD) shown in [ALI 98], with a relatively low computational effort. The computational time for the completed simulation is around 45 min on a 2.53 GHz Intel Core 2 Duo processor.

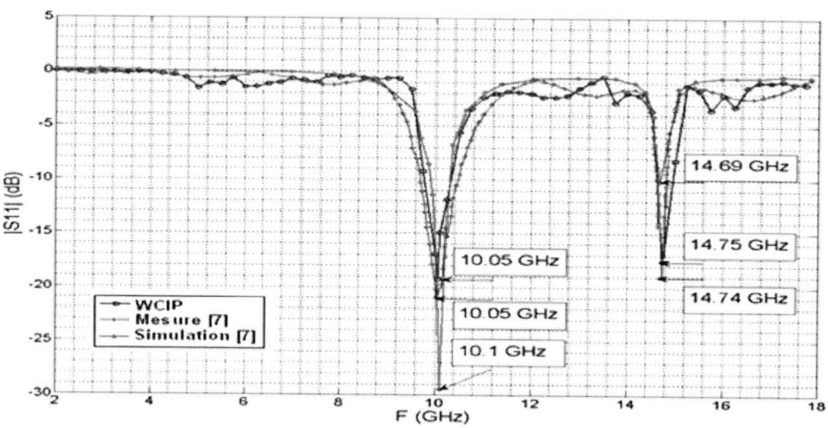

Figure 2.11. *Absolute value of the input reflection coefficient, $|S_{11}|$ in dB expressed as a function of frequency in GHz. For a color version of this figure, see www.iste.co.uk/baudrand/waves.zip*

2.3. Real and ideal polarizers within planar structures using WCIP

2.3.1. *Formulation*

An ideal polarizer, O_x, neither disrupts nor facilitates the propagation of the electric field component E_x, while it produces a complete reflexion of the component of the electric field E_y. From [2.7], it is then straightforward to express the equation E_y:

$$E_{iy} = \sqrt{Z_{0i}}\left(A_{iy} - B_{iy}\right) \qquad [2.9]$$

To meet the boundary conditions, $E_{iy} = 0$, the relationship between incident and reflective waves then becomes:

$$A_{iy} = -B_{iy} \qquad [2.10]$$

Figure 2.12 shows the physical representation of a real polarizer O_x.

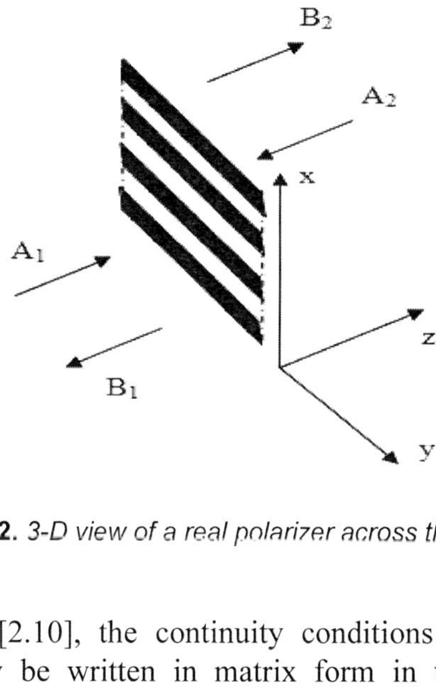

Figure 2.12. *3-D view of a real polarizer across the axis x, O_x*

Using equation [2.10], the continuity conditions across the interface in wave terms may be written in matrix form in the spatial sphere as follows:

$$\begin{bmatrix} A_{1x} \\ A_{1y} \\ A_{2x} \\ A_{2y} \end{bmatrix} = \begin{bmatrix} 0 & 0 & 1 & 0 \\ 0 & -1 & 0 & 0 \\ 1 & 0 & 0 & 0 \\ 0 & 0 & 0 & -1 \end{bmatrix} \begin{bmatrix} B_{1x} \\ B_{1y} \\ B_{2x} \\ B_{2y} \end{bmatrix} \qquad [2.11]$$

The linearity of the Fourier transform then leads us to write [2.11] as the following equation within the spatial domain:

$$\begin{bmatrix} \tilde{A}_{1x} \\ \tilde{A}_{1y} \\ \tilde{A}_{2x} \\ \tilde{A}_{2y} \end{bmatrix} = \begin{bmatrix} 0 & 0 & 1 & 0 \\ 0 & -1 & 0 & 0 \\ 1 & 0 & 0 & 0 \\ 0 & 0 & 0 & -1 \end{bmatrix} \begin{bmatrix} \tilde{B}_{1x} \\ \tilde{B}_{1y} \\ \tilde{B}_{2x} \\ \tilde{B}_{2y} \end{bmatrix} \quad [2.12]$$

The Fast Modal Transform (FMT) is made up of the algorithm 2D-FFT and the transformation from the spectral sphere to the modal sphere. This was shown in [SER 10] by using periodic walls.

If P is the FMT process and P^{-1} its converse, the incident and reflected waves may be written in the modal sphere as follows:

$$\begin{bmatrix} A_1^{TE}(k_x,k_y) \\ A_1^{TM}(k_x,k_y) \\ A_2^{TE}(k_x,k_y) \\ A_2^{TM}(k_x,k_y) \end{bmatrix} = \begin{bmatrix} P\begin{bmatrix} 0 & 0 \\ 0 & -1 \end{bmatrix}P^{-1} & P\begin{bmatrix} 1 & 0 \\ 0 & 0 \end{bmatrix}P^{-1} \\ P\begin{bmatrix} 1 & 0 \\ 0 & 0 \end{bmatrix}P^{-1} & P\begin{bmatrix} 0 & 0 \\ 0 & -1 \end{bmatrix}P^{-1} \end{bmatrix} \begin{bmatrix} B_1^{TE}(k_x,k_y) \\ B_1^{TM}(k_x,k_y) \\ B_2^{TE}(k_x,k_y) \\ B_2^{TM}(k_x,k_y) \end{bmatrix} \quad [2.13]$$

Finally, the ideal polarizer matrix may be written in the modal sphere as:

$$\begin{bmatrix} A_1^{TE}(k_x,k_y) \\ A_1^{TM}(k_x,k_y) \\ A_2^{TE}(k_x,k_y) \\ A_2^{TM}(k_x,k_y) \end{bmatrix} = \frac{1}{k_x^2+k_y^2} \begin{bmatrix} k_x^2 & k_xk_y & k_y^2 & -k_xk_y \\ -k_xk_y & -k_y^2 & k_xk_y & -k_x^2 \\ k_y^2 & -k_xk_y & k_x^2 & k_xk_y \\ k_xk_y & -k_x^2 & -k_xk_y & -k_y^2 \end{bmatrix} \begin{bmatrix} B_1^{TE}(k_x,k_y) \\ B_1^{TM}(k_x,k_y) \\ B_2^{TE}(k_x,k_y) \\ B_2^{TM}(k_x,k_y) \end{bmatrix} \quad [2.14]$$

where $k_x = (m\pi)/w_x$, $k_y = (n\pi)/w_y$, w_x and w_y are the dimensions for waveguides.

Expressing the polarizers directly within the modal sphere avoids discretization within the spatial sphere, and transformation operations with the help of the FMT and its converse. As a consequence, the computation time for these operations is shorter.

2.3.2. *Results*

To evaluate the performance of the proposed tool, three case studies have been completed. The first structure is a patch fed by a microstrip line. The other cases are real and synthetic polarizers which are introduced between the patch and the microstrip line. The wave guide is 16 mm wide and 32 mm long in dimensions, in respect of all simulations carried out.

2.3.2.1. *A multilayer structure without polarizer*

A circuit made up of two interfaces was simulated using the Matlab software program. The interfaces were discretized following the procedure described for the FMT and its related inverse. Figure 2.13(a) shows the multilayer structure with the different interfaces with $\varepsilon_{r1} = \varepsilon_{r2} = 2.2$, $\varepsilon_{r3} = 1$ and $h_1 = h_2 = 0.65$ mm. On the first interface (Figure 2.13(b)), the microstrip line has a width of 2 mm and a length of 18 mm. On the second interface (Figure 2.13(c)), the resonator has a width of 4 mm and a length of 20 mm. Figure 2.13(d) shows the anticipated behavior of the current density at a 5 GHz resonance frequency.

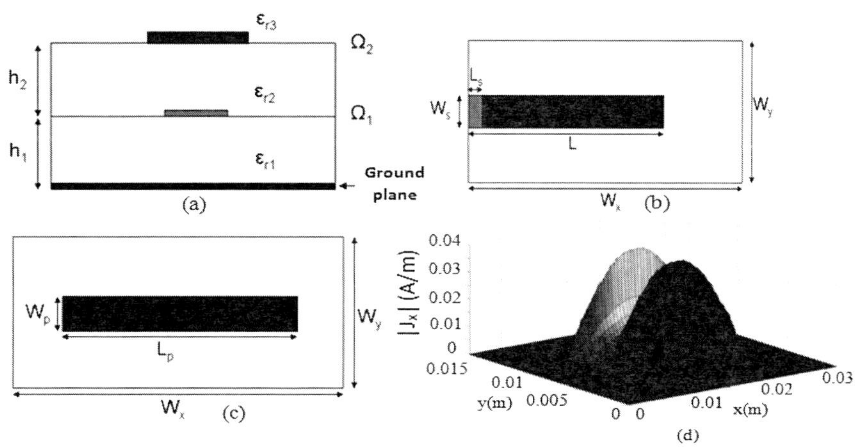

Figure 2.13. *a) Multilayer structure, line; b) the feed line across the first interface; c) the resonator on the second interface; d) behavior of current density across the second interface at 5 GHz*

2.3.2.2. *Multilayer structure with a real polarizer*

In this case, a second interface corresponding to the real polarizer was placed between the power supply and the resonator (Figure 2.14(a)). The physical and electric parameters are: $\varepsilon_{r1} = \varepsilon_{r2} = \varepsilon_{r3} = 2.2$, $\varepsilon_{r4} = 1$, $h_1 = 0.65$ mm, $h_2 = 0.3$ mm and $h_3 = 0.35$ mm.

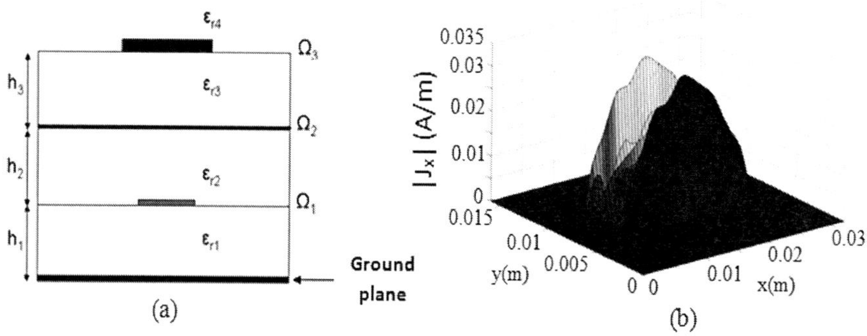

Figure 2.14. *a) Multilayer structure; b) behavior of the current density across the third interface at 4.1 GHz*

The polarizer is made up of parallel metal strips in the direction *Oy*. These have a width of 2 mm and a length of 32 mm spaced at intervals of 2 mm. For this simulation, the three interfaces are discretized with all operations of the FMT and its linked converse. Figure 2.14(b) shows the behavior of current density across the resonator with the resonance frequency of 4.1 GHz. This new resonance frequency is due to the increase in capacitance and inductance of the real polarizer. In Figure 2.14(b), the influence of the metallic parallel strips upon the behavior of current density may be seen.

2.3.2.3. *Multilayer structure with ideal polarizer*

Figure 2.15(a) shows the same structure, where the second interface is replaced by the ideal polarizer model. In this case, only two interfaces (the power line and the resonator) are discretized together with all FMT calculations and its linked converse. The ideal polarizer interface is considered as virtual with all calculations taking place only in the modal sphere.

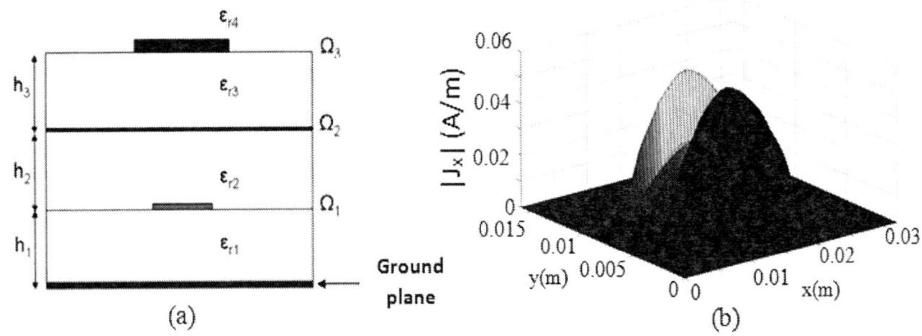

Figure 2.15. *a) Multilayer structure; b) behavior of current density across the third interface at 3.9 GHz*

Figure 2.15(b) shows the behavior of current density across a resonator with a resonance frequency at 3.9 GHz. This resonance frequency is close to that obtained for the real polarizer without the influence of parallel metallic strips. Moreover, a saving of 30% in the computation time was observed.

2.4. Amplifier structure of compact micro-waves

In this section, the analysis and simulation of a five-interface structure shown on Figure 2.16 was completed.

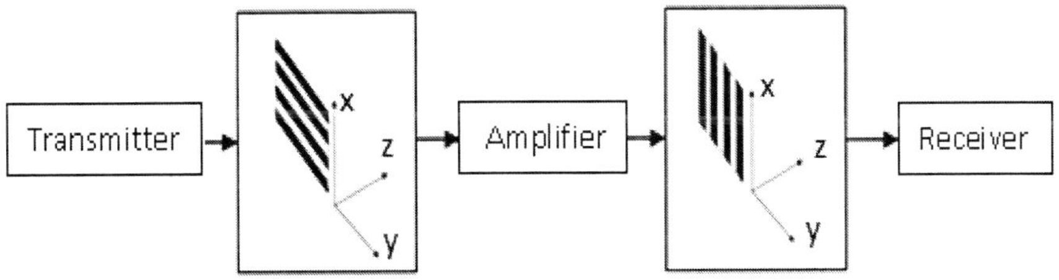

Figure 2.16. *Block diagram of an amplifier*

2.4.1. *Formulation of the amplifier interface*

To model the amplifier interface, the auxiliary sources method is used. This method was demonstrated by formula [SBO 01] and, more recently, through using formula [HAR 13]. It is specifically adapted to the electromagnetic analysis of active elements for the planar circuit. This method allows for electromagnetic modeling of the structure made up of passive elements (P) and an active element (A), as shown in Figure 2.17.

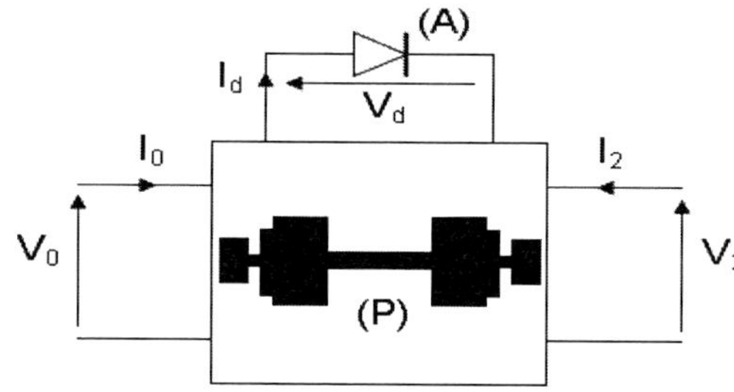

Figure 2.17. *Example of an active circuit*

The first stage of the method is electromagnetic analysis of the passive structure by replacing the active element by an auxiliary source as shown in Figure 2.18(a).

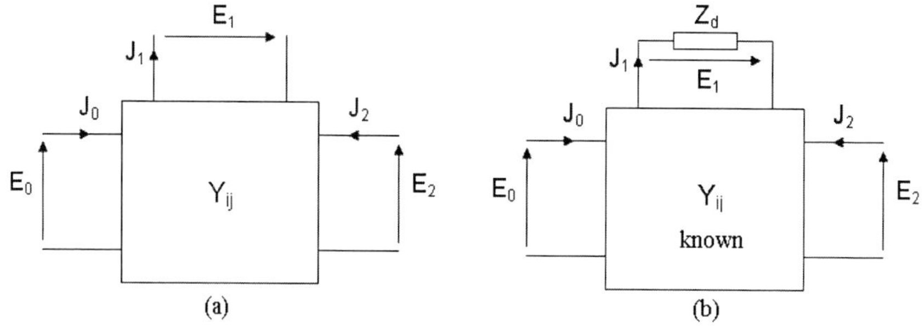

Figure 2.18. *a) Analysis of the passive structure; b) analysis of the active structure*

Using the WCIP, it is possible to obtain the Y_{ij} admittance matrix of the three-port circuit structure applying the formula below:

$$\begin{bmatrix} J_0 \\ J_1 \\ J_2 \end{bmatrix} = \begin{bmatrix} Y_{11} & Y_{12} & Y_{13} \\ Y_{21} & Y_{22} & Y_{23} \\ Y_{31} & Y_{32} & Y_{33} \end{bmatrix} \begin{bmatrix} E_0 \\ E_1 \\ E_2 \end{bmatrix} \qquad [2.15]$$

The second stage consists of substituting the auxiliary source, in this case, E_1, with the electric representation of the active element. This is illustrated in Figure 2.5(b), where Z_d represents the impedance of a diode. Considering that this active element is a localized element with a negative impedance [TRA 03] the relevant equation is as follows:

$$\begin{aligned} J_0 &= Y_{11}E_0 + Y_{12}E_1 + Y_{13}E_2 \\ J_1 &= Y_{21}E_0 + Y_{22}E_1 + Y_{23}E_2 = -\frac{E_1}{Z_d} \\ J_2 &= Y_{31}E_0 + Y_{32}E_1 + Y_{33}E_2 \end{aligned} \quad [2.16]$$

A new admittance matrix is computed. The hexapole becomes a two-port circuit network.

$$\begin{cases} J_0 = Y_{11Ampl}E_0 + Y_{12Ampl}E_2 \\ J_2 = Y_{21Ampl}E_0 + Y_{22Ampl}E_2 \end{cases} \quad [2.17]$$

The coefficients of the equivalent diffraction matrix are then computed from the admittance matrix of this structure:

$$S_{11} = \frac{(1 - Z_c Y_{11Ampl})(1 - Z_c Y_{22Ampl}) + Z_c^2 Y_{12Ampl} Y_{21Ampl}}{(1 + Z_c Y_{11Ampl})(1 - Z_c Y_{22Ampl}) - Z_c^2 Y_{12Ampl} Y_{21Ampl}} \quad [2.18]$$

$$S_{21} = \frac{-2 Z_c Y_{21Ampl}}{(1 + Z_c Y_{11Ampl})(1 - Z_c Y_{22Ampl}) - Z_c^2 Y_{12Ampl} Y_{21Ampl}} \quad [2.19]$$

where Z_c is the typical impedance of 50 Ω.

2.4.2. *The simulation results*

To set out and evaluate the suggested structure, a number of simulations of the amplifying structure have been completed. Figure 2.19 shows the multilayer structure with the physical interface Ω_2 used for the auxiliary source method.

On the first interface, the microstrip power line has a width of 2 mm and a length of 18 mm. On the second interface the resonator in *L* has the dimensions 4 mm x 20 mm (*l* x *L*). On the third interface the microstrip line has the same dimensions as the first interface. Expressing the polarizers directly within the modal sphere avoids discretization within the spatial domain. Consequently, 30% of computation time is saved. The dimensions of the wave guide are 32 mm x 32 mm, with $\varepsilon_{r1} = \varepsilon_{r2} = \varepsilon_{r3} = 2.2$, $\varepsilon_{r4} = 1$. The last layer (the upper layer) is free space, and $l_1 = l_2 = l_3 = 0.65$ mm.

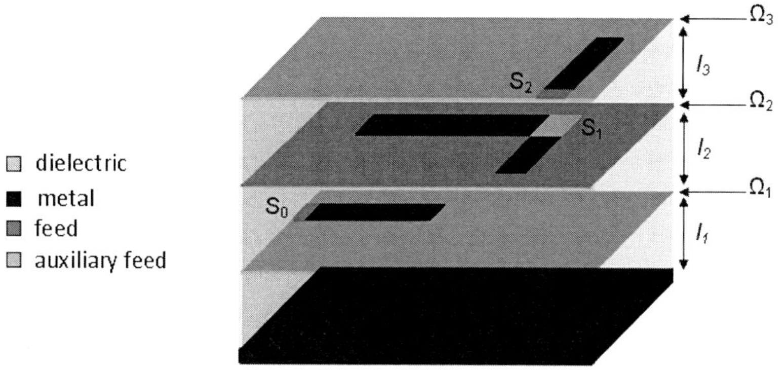

Figure 2.19. *3-D view of the auxiliary feed structure. For a color version of this figure, www.iste.co.uk/baudrand/waves.zip*

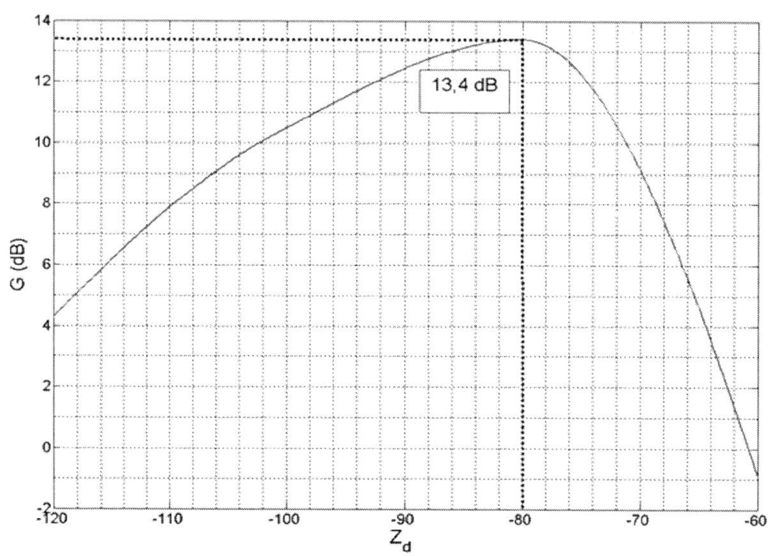

Figure 2.20. *Amplification according to the impedance of the amplifying structure*

The simulations of the suggested structure have been achieved by varying the negative impedance of the diode from –120 Ω to –60 Ω for a 5.6 GHz frequency. As Figure 2.20 shows, a maximum amplification of 13.4 dB is obtained with a Z_d impedance of –80 Ω.

3

Applications of the WCIP Method to Frequency Selective Surfaces (FSS)

The iterative method Wave Concept Iterative Procedure (WCIP) as applied to planar structures is set out. Based upon a formulation in transverse waves and wave gathering at interface level, the solution to the electromagnetic issue is obtained through using an iterative procedure. The use of bounded operators within the iterative method makes convergence a certainty, independently of the planar circuit on the frequency selective surfaces (FSS) interfaces. Diodes, inserted to obtain an active FSS structure, as well as gaps and short-circuits have the role of both controlling and adjusting the characteristics of FSS diffractions. Within Frequency Selective Surfaces (FSSs) controlled by diodes, the latter are considered surface impedances within the iterative method and its two-port network may thus be established. The WCIP method, switches between the spatial domain and the modal domain via the Fast Modal Transform (FMT) – to resolve the electromagnetic issue. The FMT based upon the algorithm of the bi-dimensional FFT (Fast Fourier Transform) then allows significant savings both in terms of computational time and memory space. Several FSS structures have been analyzed by using the WCIP method. The results obtained are compared with the measurement results and the simulation from the works in the field, and an overall consensus of findings is then evident.

Chapter written by Mohammed TITAOUINE and Henri BAUDRAND.

3.1. Introduction

Frequency Selective Surfaces (FSS) are largely used as low-pass filters, high-pass filters, notch filters, or band-pass filters [GAO 05, MON 05]. Charged Frequency Selective Surfaces (FSSs) make flexibility within the adjustment of transmission characteristics possible. However, electronically-adjusted, or so-called active FSSs, are very significant in compensating for manufacturing errors and for the adaptation of the FSS filters to frequency variations in the operation of wireless communication systems [MAR 05, MAR 05]. The use of short-circuits within FSS dielectric rings ensures precise resonance frequency control [MAR 01] by varying the position of inserted short-circuits. For multi-band and large-band, multi-ring FSSs [VOY 06, AUB 06, QUI 06] and multilayer plated FSSs are used [UCA 08, SON 08, ERD 08]. It is possible to create dual-polarized FSSs when a careful decision upon FSS circuit is made.

Several methods have been used to resolve such problems. Amongst these methods, the Finite-Difference-Time-Domain (FDTD) method, the Transmission Line Matrix (TLM), the immittance method in conjunction with the method of moments [CAM 02, GOM 02] are currently adopted. Nevertheless, the tendency towards "Full Wave" simulations of entire microwave systems is expanding rapidly [WAN 03, BAU 03]. Indeed, this approach gives more accurate results, when compared with multi-block modeling and assembly with a series of element behaviors, making up a comprehensive circuit, where the linking of elements is simply neglected. The iterative "Wave Concept Iterative Procedure" (WCIP) method, developed for several years within the electronics laboratory LEN7 in Toulouse France, responds to these demands and offers an efficient and general solution for planar structures, when the analysis revolves around transverse waves. It also sets out a solution for all types of structure if volumes waves are used.

The iterative WCIP method considers the FSS structure as a form of periodic structure. The analysis of such a structure is reduced to that of the unit cell, indeed a so-called "repeated cell", to determine characteristics of the entire FSS structure. Virtual walls, so-called periodic walls, are assumed to separate unit cells within the FSS structure. The analysis using the iterative WCIP method is completed in two essential, but repeated, stages until method convergence happens. The first stage is the diffraction of incident waves across the interface, which is characterized by the presence of metal, and the second stage is the wave reflection by the covers enclosing the unit cell of the given FSS, which is subject to analysis. The projection of diffracted waves within the base of modes of periodic wall housing (whether TEM modes – transverse electromagnetic modes, TE modes – transverse electric modes or TM modes – transverse magnetic modes) takes place by using FMT-Fast modal Transform. In returning to the spatial domain, the inverse transform applying the FMT^{-1} mode is used. The use of the algorithm bi-dimensional FFT algorithm makes savings possible both in terms of computation time and the required memory space.

The results obtained via the iterative WCIP method accord well with the measurement results, carried out at the Centre Fédéral d'Éducation Technologique de Paraíba João Pessoa du Brésil (a technological education institution located in João Pessoa, Paraíba, Brazil), together with the results of field findings.

3.2. Formulation of the iterative WCIP method

The iterative WCIP is based around the formulation in transverse waves and the information gathering at the interface, which contains the given circuit. The iterative method is summarized in two essential stages. The first stage represents the diffraction of the incident wave across the interface and the second stage the reflected waves by the closure covers of the rectangular housing with periodic walls. Hence, the use of two given operators: the diffraction operator, defined within the spatial domain, and the reflective operator defined with the spectral domain. These two operators are of a restricted type, hence the method convergence being ensured, independently of the structure studied.

FSSs are periodic structures, operating as filters and displaying spectral selectivity, which depend upon the incident wave polarization, the geometry of the planar circuit and the separation distance between the FSS elements.

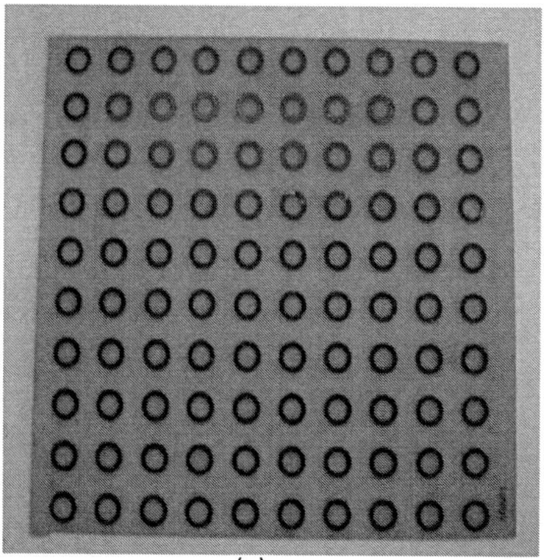
(a)

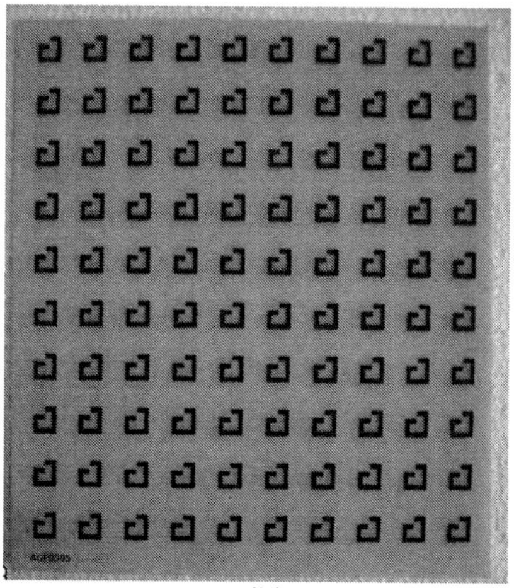
(b)

Figure 3.1. *Some examples of FSS; a) FSS examples with circular metallic rings, and b) quasi-open square rings*

To explain the WCIP method, the FSS structure in Figure 3.2 may be studied.

In the iterative WCIP method, the FSS structure is regarded as a periodic structure and the analysis thereof breaks down to to the unit cell shown in Figure 3.2(b). The broken lines in Figure 3.2(a) represent virtual periodic walls, which are assumed to separate adjacent FSS cells. The dual media issue within the iterative method is shown in Figure 3.1(c) where the waves are on both sides of the interface Ω on which the FSS circuit is etched.

Incident waves $\vec{A}_i$ and diffracted waves $\vec{B}_i$ within the media i are calculated from the tangential fields by [RAV 03, CON 99] using the following equations:

$$\begin{aligned} \vec{A}_i &= \frac{1}{2\sqrt{Z_{0i}}}\left(\vec{E}_i + Z_{0i}\,\vec{J}_i\right) \\ \vec{B}_i &= \frac{1}{2\sqrt{Z_{0i}}}\left(\vec{E}_i - Z_{0i}\,\vec{J}_i\right) \end{aligned} \quad [3.1]$$

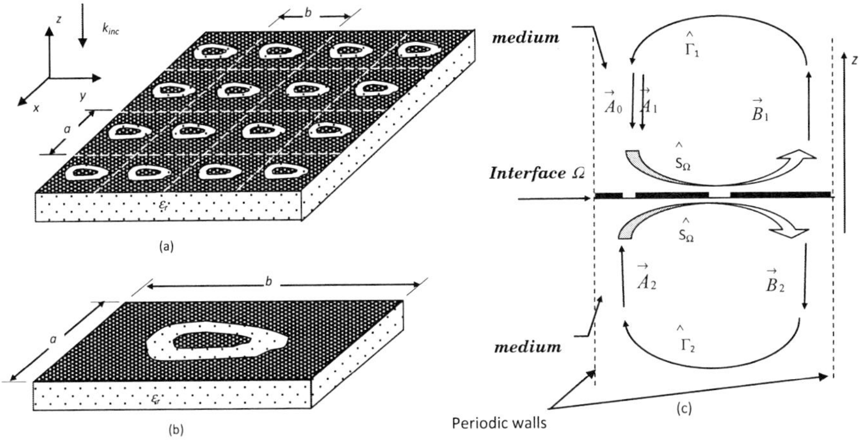

Figure 3.2. *a) Geometry of an FSS with dielectric rings taking an arbitrary shape; b) unit cell within the FSS structure; c) section of the unit cell of; b) along the y axis*

Z_{oi} is the typical impedance of the medium *i*. The current density $\vec{J}_i$ of the medium *i* is linked to the magnetic field by the equation:

$$\vec{J}_i = \vec{H}_i \wedge \vec{n} \quad [3.2]$$

where $\vec{n}$ is the normal vector of the interface Ω. The electric field and the current density may be calculated from the transverse waves by the equation:

$$\vec{E}_i = \sqrt{Z_{oi}}\left(\vec{A}_i + \vec{B}_i\right) \quad [3.3a]$$

$$\vec{J}_i = \frac{1}{\sqrt{Z_{oi}}}\left(\vec{A}_i - \vec{B}_i\right) \quad [3.3b]$$

In referring to Figure 3.2, diffracted waves $\vec{B}_i$ are obtained from incident waves $\vec{A}_i$ by applying the diffraction operator $\hat{S}_\Omega$ with the following equation:

$$\begin{vmatrix} \vec{B}_1 \\ \vec{B}_2 \end{vmatrix} = \hat{S}_\Omega \begin{vmatrix} \vec{A}_1 \\ \vec{A}_2 \end{vmatrix} \quad [3.4]$$

The diffracted waves $\vec{B}_i$ are reflected across the housing casing covers with periodic walls enclosing the FSS unit cell, and becoming incident waves $\vec{A}_i$ after the addition of the source $\vec{A}_0$. The application of the reflection operator to diffracted waves is given in the equation:

$$\begin{vmatrix} \vec{A}_1 \\ \vec{A}_2 \end{vmatrix} = \hat{\Gamma} \begin{vmatrix} \vec{B}_1 \\ \vec{B}_2 \end{vmatrix} + \begin{vmatrix} \vec{A}_0 \\ 0 \end{vmatrix} \qquad [3.5]$$

$\hat{\Gamma}_i$ is the reflection coefficient.

A flow diagram of the WCIP method, applied to FSS, is given in Figure 3.3.

3.2.1. *Determining the diffraction operator*

The diffraction operator is determined from both the FSS geometry and the interface boundary conditions Ω.

Two types of FSS may be considered. Those FSSs which are uncharged and those which are charged. Charges are inserted within the FSS structure to ensure a better flexibility within the FSS selectivity. Generally, there are three domains: the metallic domain, the dielectric domain and the electrical charge domain. Uncharged FSSs are obtained by associating the charge domain with the dielectric domain or the metallic domain. The charge domain may contain several lumped elements (for example resistances, capacitances, inductances). These three domains are defined by Heaviside's functions, that is to say the equations:

$$H_I = \begin{cases} 1, & on\ the\ dielectric \\ 0, & otherwise \end{cases} \qquad [3.6a]$$

$$H_M = \begin{cases} 1, & on\ the\ metal \\ 0, & otherwise \end{cases} \qquad [3.6b]$$

$$H_Z = \begin{cases} 1, & on\ \ the\ \ surfaces\ occupied\ by\ the\ loads \\ 0, & otherwise \end{cases} \qquad [3.6c]$$

The boundary conditions, at the interface level, within each domain are:

$$\vec{E}_1 = \vec{E}_2 = \vec{0} \qquad [3.7]$$

within the metallic domain H_M.

$$\begin{cases} \vec{J}_{tot} = \vec{J}_1 + \vec{J}_2 = \vec{0} \\ \vec{E}_1 = \vec{E}_2 \end{cases} \qquad [3.8]$$

within the (dielectric) insulating domain H_I.

In the electrical charge domain, the boundary condition is shown by the formula:

$$\begin{cases} \vec{E}_1 = Z_s \vec{J}_{tot} \\ \vec{E}_2 = Z_s \vec{J}_{tot} \end{cases}, \quad \vec{J}_{tot} = \vec{J}_1 + \vec{J}_2 \qquad [3.9]$$

The impedance of the surface Z_s is calculated from the impedance Z_L from the lumped element, using the formula:

$$Z_s = \frac{W_z}{L_z} Z_L \qquad [3.10]$$

where W_z and L_z are the width and length of the equivalent domain of the lumped element.

By applying the boundary conditions to the interface Ω and upon each domain, the global diffraction operator may be calculated by the equations:

$$\left[\hat{S}_\Omega\right] = \begin{vmatrix} -H_M + \dfrac{1-N^2}{1+N^2} H_I + \dfrac{-Z_{01}Z_{02} - Z_{01}Z_s + Z_{02}Z_s}{Z_{01}Z_{02} + Z_{01}Z_s + Z_{02}Z_s} H_z & \dfrac{2N}{1+N^2} H_I + \dfrac{2Z_s\sqrt{Z_{01}Z_{02}}}{Z_{01}Z_{02} + Z_{01}Z_s + Z_{02}Z_s} H_z \\ \dfrac{2N}{1+N^2} H_I + \dfrac{2Z_s\sqrt{Z_{01}Z_{02}}}{Z_{01}Z_{02} + Z_{01}Z_s + Z_{02}Z_s} H_z & -H_M - \dfrac{1-N^2}{1+N^2} H_I + \dfrac{-Z_{01}Z_{02} + Z_{01}Z_s - Z_{02}Z_s}{Z_{01}Z_{02} + Z_{01}Z_s + Z_{02}Z_s} H_z \end{vmatrix}$$

[3.11]

with $N = \sqrt{Z_{o1}/Z_{O2}}$.

3.2.2. Determining the reflection operator

Not knowing the reflection coefficient with the spatial domain, the transition to the modal domain is essential. As a consequence, the projection of diffracted waves on the modal basis must occur before the calculation of the reflected waves. In the case of FSSs, the entire modal basis is made up of the TEM mode, the TE_{mn} modes and the TM_{mn} modes. It is a question of the transition from the spatial domain to the modal domain, which is achieved through using the FMT. When reflected waves are obtained and the source addition is then made, it is necessary to return to the spatial domain via the use of the inverse FMT.

The reflection coefficient within the modal domain is given as the equation:

$$\Gamma i_{mn}^{\alpha} = \frac{1 - Z_{oi} Yi_{mn}^{\alpha}}{1 + Z_{oi} Yi_{mn}^{\alpha}} \qquad [3.12]$$

where Yi_{mn}^{α} is the admittance of the mnth mode at the medium i and α represents the TE or TM modes. Here Yi_{mn}^{α} may be calculated by:

$$\hat{Y}i_{mn}^{TM} = \frac{j\omega\varepsilon_o\varepsilon_{ri}}{\overset{(i)}{\gamma}_{mn}} \qquad \hat{Y}i_{mn}^{TE} = \frac{\overset{(i)}{\gamma}_{mn}}{j\omega\mu_o} \qquad [3.13]$$

$\overset{(i)}{\gamma}_{mn}$ is the propagation constant of the mode mn within the medium i. It may be obtained by the equation:

$$\overset{(i)}{\gamma}_{mn} = \sqrt{\left(\frac{2\pi m}{a}\right)^2 + \left(\frac{2\pi n}{b}\right)^2 - k_o^2\varepsilon_{ri}} \qquad [3.14]$$

ε_o, ε_{ri} and μ_o are, respectively, the permittivity in a vacuum, the permittivity relating to the medium i and the permeability of the vacuum.

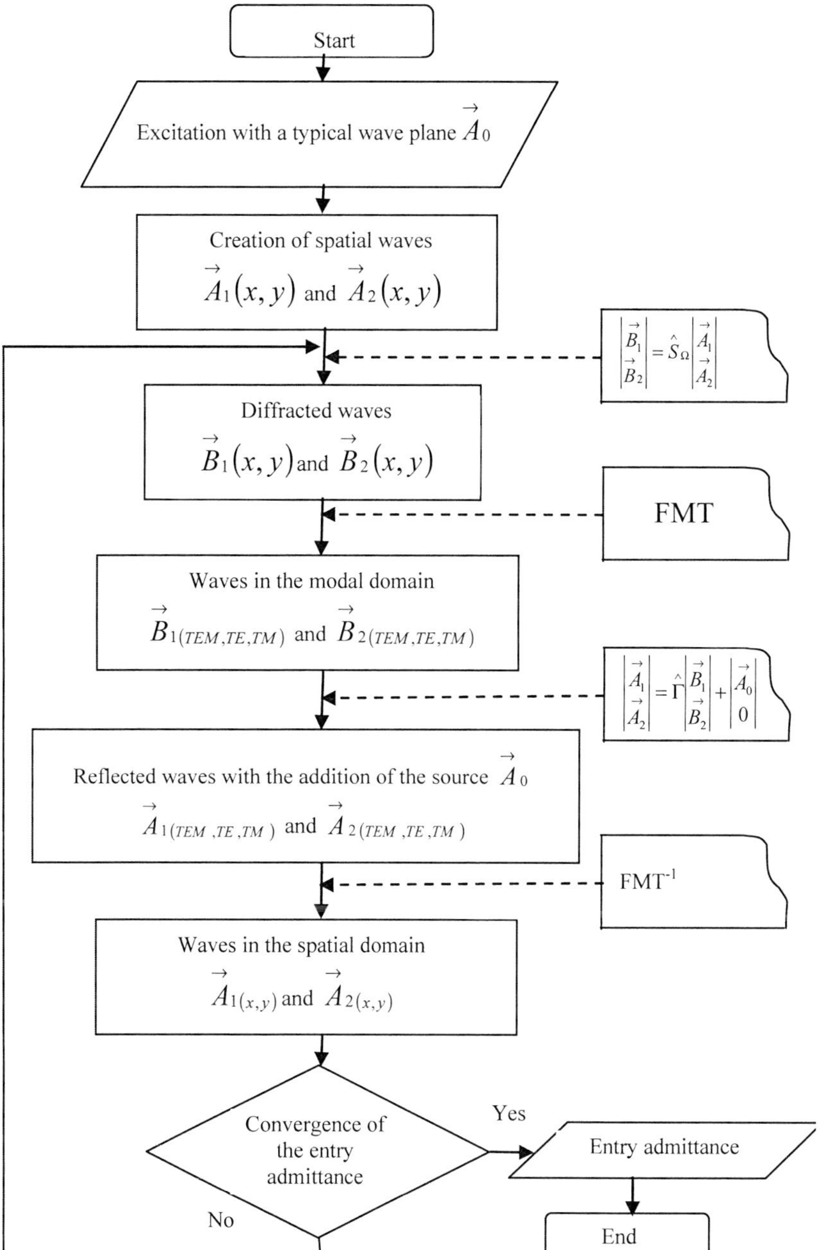

Figure 3.3. *Flow diagram of the WCIP method*

3.2.3. *The fast modal transform FMT and its inverse FMT^{-1}*

The FMT facilitates the transition from the spatial domain to the modal domain, and the inverse FMT enables the return to the spatial domain with the following relevant equations:

$$\begin{aligned} FMT\left(\begin{vmatrix} E_x(x,y) \\ E_y(x,y) \end{vmatrix}\right) &= \begin{vmatrix} B_{mn}^{TE} \\ B_{mn}^{TM} \end{vmatrix} = \begin{vmatrix} K_{ymn} & -K_{xmn} \\ K_{xmn} & K_{ymn} \end{vmatrix} FFT2\left(\begin{vmatrix} E_x(x,y) \\ E_y(x,y) \end{vmatrix}\right) \\ FMT^{-1}\left(\begin{vmatrix} B_{mn}^{TE} \\ B_{mn}^{TM} \end{vmatrix}\right) &= \begin{vmatrix} E_x(x,y) \\ E_y(x,y) \end{vmatrix} = FFT2^{-1}\left(\begin{vmatrix} K_{ymn} & K_{xmn} \\ -K_{xmn} & K_{ymn} \end{vmatrix}\begin{vmatrix} B_{mn}^{TE} \\ B_{mn}^{TM} \end{vmatrix}\right) \end{aligned} \quad [3.15]$$

The logical development of [3.15] is set out in [TIT 07a].

3.2.4. *FSS multilayer devices*

The issue of multilayer FSSs is considered in an inherent way within the analysis of FSSs on a single interface, since the air layer adds to the lower layer and the problem is reduced to being simply the WCIP issue with two media (Figure 3.2) by bringing the admittance of the air layer to the interface for each mode. Generally, multilayer FSSs are dealt with by bringing the admittance of the air layer to the interface Ω, covering the various dielectric layers and using the equation:

$$Y_{(i)mn\,\Omega}^{\alpha} = Y_{(i)mn}^{\alpha}\left(\frac{Y_{(i-1)mn}^{\alpha} + Y_{(i)mn}^{\alpha}\tanh\left(\overset{(i)}{\gamma}_{mn} h_i\right)}{Y_{(i)mn}^{\alpha} + Y_{(i-1)mn}^{\alpha}\tanh\left(\overset{(i)}{\gamma}_{mn} h_i\right)}\right) \quad [3.16]$$

Thus, the admittance $Y_{(i-1)mn}^{\alpha}$ is brought back to the interface separating both layers (i+1) and i through the ith layer thickness h_i using $Y_{(i)mn}^{\alpha}$ as the ith layer characteristic impedance for the α mode.

3.2.5. *Multi-level plated FSSs*

Multi-level plating FSSs may be used to achieve multi-frequency band or broadband FSSs. Within the iterative method, the layer separating two consecutive interfaces is modeled as a guide length, and a transition operator between these two layers is defined within the modal domain and is an intermediate stage within the calculation. This operator is given by [GAR 01] with the equation as follows:

$$\left[\hat{S}_{hi}\right] = \begin{vmatrix} \dfrac{\left(Z_{mn}^{\alpha\,2} - Z_{0i}^{2}\right)\sinh\left(\gamma_{mn}h_i\right)}{\Delta} & \dfrac{2Z_{mn}^{\alpha}Z_{0i}}{\Delta} \\ \dfrac{2Z_{mn}^{\alpha}Z_{0i}}{\Delta} & \dfrac{\left(Z_{mn}^{\alpha\,2} - Z_{0i}^{2}\right)\sinh\left(\gamma_{mn}h_i\right)}{\Delta} \end{vmatrix} \qquad [3.17]$$

with $\Delta = 2Z_{mn}^{\alpha}Z_{0i}\cosh\left(\gamma_{mn}h_i\right) + \left(Z_{mn}^{\alpha\,2} + Z_{0i}^{2}\right)\sinh\left(\gamma_{mn}h_i\right)$

h_i is the thickness of the layer separating both consecutive interfaces and $Z_{(i)mn}^{\alpha}$ is the impedance of the mode mn of the type α, which may be TE or TM.

In Figure 3.4, an example of an FSS double-plated layer and its corresponding diagram are shown.

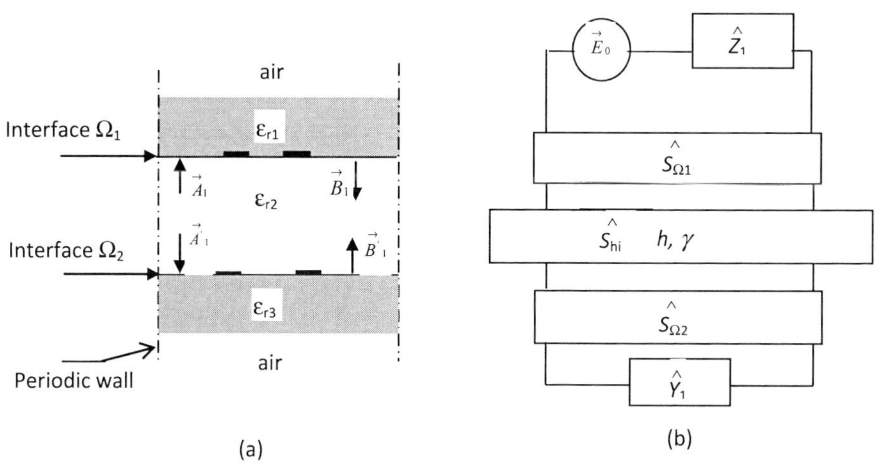

Figure 3.4. *FSS double-layer plating a) section of unitary cell; b) corresponding diagram*

3.3. Application of the iterative WCIP method to different FSSs

The iterative method is applied to several types of FSS such as the dielectric short-circuited ring FSSs [TIT 08], the FSS charged with lumped elements and the active FSS [TIT 07], FSS concentric metallic rings [TIT 08], FSS double polarization and FSS broadband. The results obtained with the WCIP method are shown with both the measurement results and/or simulation measurement results to enable a comparison.

3.3.1. *Dielectric short-circuited FSS rings*

The insertion of two short-circuits and four short-circuits in variable angular positions offers more flexibility in the adjustment of the FSS resonance frequency with circular dielectric rings shown in Figure 3.5. Circular rings are printed on a thickness substrate of 0.51 mm and a dielectric constant of 3.4. The internal radius and the external radius of the dielectric rings are respectively 3.8 mm and 4.5 mm. The dimensions of the unitary cell are a = 11.43 mm and b = 10.13 mm.

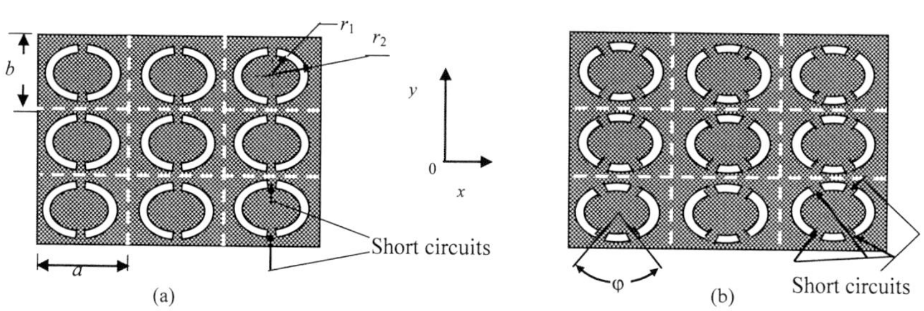

Figure 3.5. *Circular ring FSS with short circuits, a) FSS with two short circuits; b) FSS with four short circuits*

Figure 3.6 shows the reflection coefficient of the FSS in Figure 3.5 when it is fed with a normally incident plane wave in both the x and the y polarizations, respectively. The interface is described as having a grid of 120 × 120 pixels and the iterative procedure is halted after 1,600 iterations.

For increased flexibility within the resonance frequency, two other short-circuits are added in specific angular positions as in Figure 3.5(b). Figure 3.7

shows the reflection coefficient according to the frequency for $\varphi = 20°$, $40°$, $30°$ and $60°$, when the FSS is illuminated by a normally incident plane wave in the y polarization.

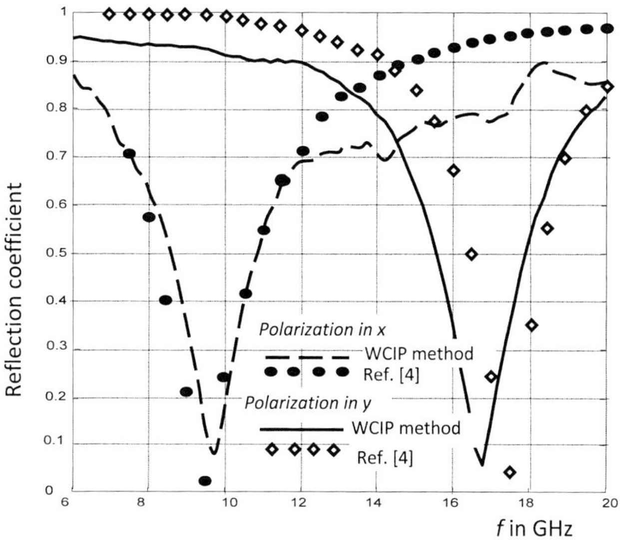

Figure 3.6. *Reflection coefficient variation according to the FSS frequency with two short ring circuits with a normally incident source in the x polarization and in the y polarization [TIT 08]*

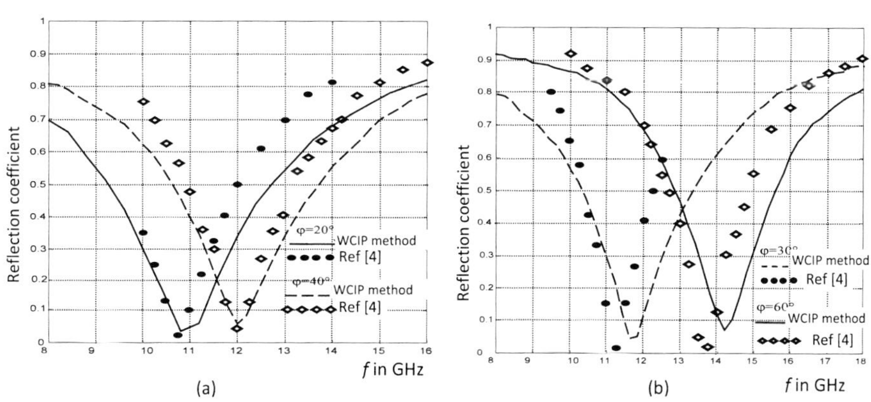

Figure 3.7. *Variation of the reflection coefficient according to frequency, with different values for the angular position φ, the four FSS circular ring short circuits with a normal incidence, that is polarized in y [TIT 08], a) $\varphi = 20°$, $\varphi = 40°$ (b) $\varphi = 30°$, $\varphi = 60°$*

The variation of the angular position of the four short-circuits $\varphi = 20°$ at $\varphi = 100°$ results in scanning a frequency band which ranges from around 10.9 GHz to 19.2 GHz [MAR 01, TIT 08].

3.3.2. *FSSs charged by lumped elements and active FSSs*

Using FSSs charged with passive elements and active elements is another alternative way to achieve the FSS concept, with a more variable selectivity. To check that the WCIP iterative method supports charged FSSs, it is applied to a circular FSS dielectric ring, which is charged by capacitors, inductances or PIN diodes.

Figure 3.8 shows the reflection coefficient of this FSS, with a thickness substrate of 0.102 mm and a dielectric constant of 2.4. It has a resonance of nearly 13.9 GHz.

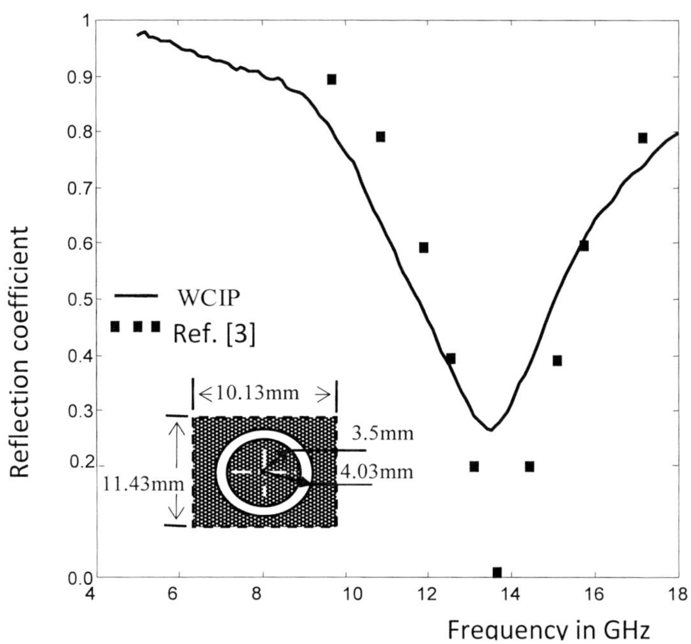

Figure 3.8. *Reflection coefficient of an uncharged dielectric FSS circular ring [TIT 08]*

3.3.2.1. *FSS charged by passive elements*

The circular FSS rings are now charged by capacitors and inductances, and corresponding reflection coefficients are shown in Figures 3.9(b) and (c), respectively.

The increase in capacitance decreases the resonance frequency in relation to an uncharged FSS. On the other hand, the increase in inductance also increases the resonance frequency of the FSS, charged with inductances, as Figure 3.9 shows. Hence, a large frequency range, centered upon the resonance frequency of an uncharged FSS, may be detected.

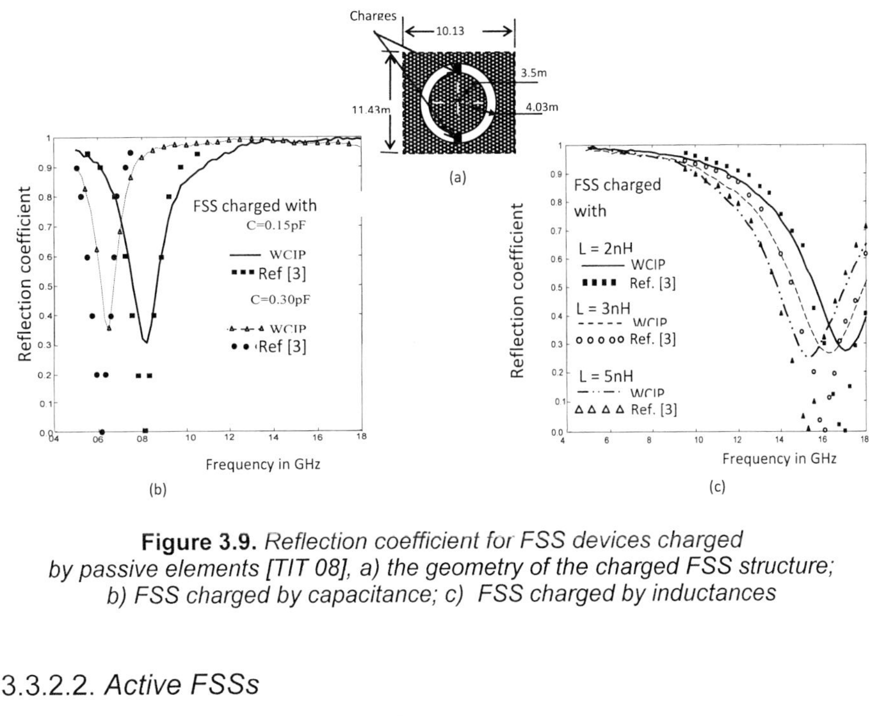

Figure 3.9. *Reflection coefficient for FSS devices charged by passive elements [TIT 08], a) the geometry of the charged FSS structure; b) FSS charged by capacitance; c) FSS charged by inductances*

3.3.2.2. *Active FSSs*

To avoid regularly changing the capacitance to vary the FSS frequency, which is charged with capacitors, PIN diodes may be used to give electronically-controlled FSS devices. Active FSS devices are highly significant in compensating for manufacturing errors and adapting the FSS filter to variations in operational frequency of wireless communication systems [TIT 07]. It is enough to vary the reverse bias voltage for the capacitance of the PIN diode to then fluctuate. The inverse polarization diodes may be modeled as a resistance connected in parallel with a capacitance [MAR 05]. As a consequence, the analysis of the active FSS is reduced to the FSS being charged by a charge which has an impedance of the surface Z_s (that being obtained from the resistance in parallel with the capacitance modeling the reverse bias diode).

Figure 3.10 shows the reflection coefficient with the same FSS structure. However this time, it is charged with PIN diodes which are polarized with a voltage of –30 V, –3 V and 0 V.

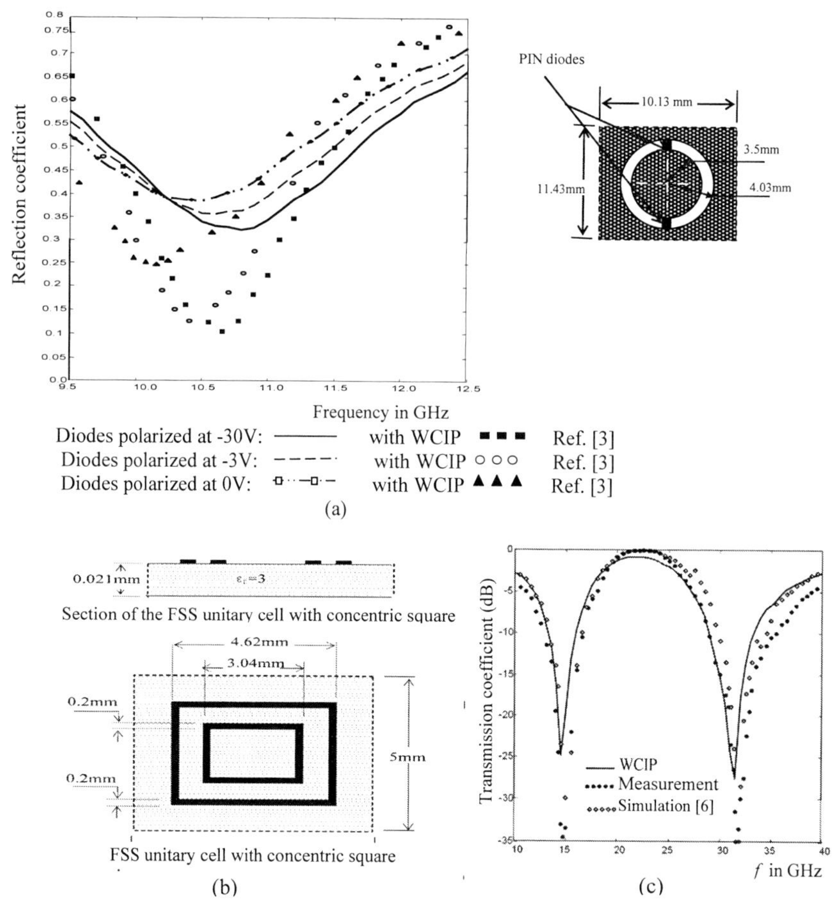

Figure 3.10. *FSS unitary cell with circular rings and the FSS with the two rectangular rings: a) FSS reflection coefficient with dielectric circular rings charged with PIN iodes [TIT 08]; b) FSS geometry with two metallic concentric discs [TIT 08]; c) variation of the transmission coefficient according to the operating frequency [TIT 08]*

3.3.3. *Multi-frequency band FSSs*

In the case of FSS single interfaces and for multi-frequency band applications, the metallic concentric FSS rings [TIT 08] or FSSs with fractal geometry [KIM 06] may be used. The open quasi-square FSSs with metallic rings allow two frequency bands in a given polarization direction, and a single frequency band in the perpendicular polarization direction.

3.3.3.1. *FSS with metallic rectangular concentric rings*

Figure 3.10(a) shows a FSS with two metallic rectangular concentric rings. The structure is fed with a normally incident plane wave. The transmission coefficient represented in Figure 3.10(b) shows the presence of a second frequency band, which is due to the addition of a second ring.

3.3.3.2. *Metallic FSS quasi-rectangular open rings*

The FSS structure with metallic open quasi-square rings is shown in Figure 3.1(b) and the unitary cell is shown in Figure 3.11(a). The FSS structure is printed on a substrate with both a thickness of 1 mm and a dielectrc constant of 4. Figure 3.11(b) shows the transmission coefficient of the FSS with metallic open quasi-square rings when it is illuminated by a normally incident wave plane in both of the polarizations x and y. This FSS offers scope for precise resonance frequency adjustment for the two frequency bands within the polarization x and that of the single band in polarization y, by varying the coupling length of the parallel FSS strips. The presence of both resonances as well as their variations is due to two strip lengths being understood to have an assumed quasi-symmetrical axis in the y direction. The resonance variation at the time of the polarization of y is caused by capacitance created by both strips, and the increase in the coupling length also increases the facing surface area of the capacitance metallic plates.

3.3.3.3. *U-shaped FSS*

U-shaped FSSs afford two rejection bands if the source is polarized in the x direction. Figure 3.12(a) shows the metallic pattern U-shaped FSS unitary cell, with dimensions 20 mm x 20 mm. In Figure 3.12(b), the transmission coefficient is shown. The two resonances are controlled by the two lengths d_1 and d_2, which are both inversely proportional to the resonance frequencies.

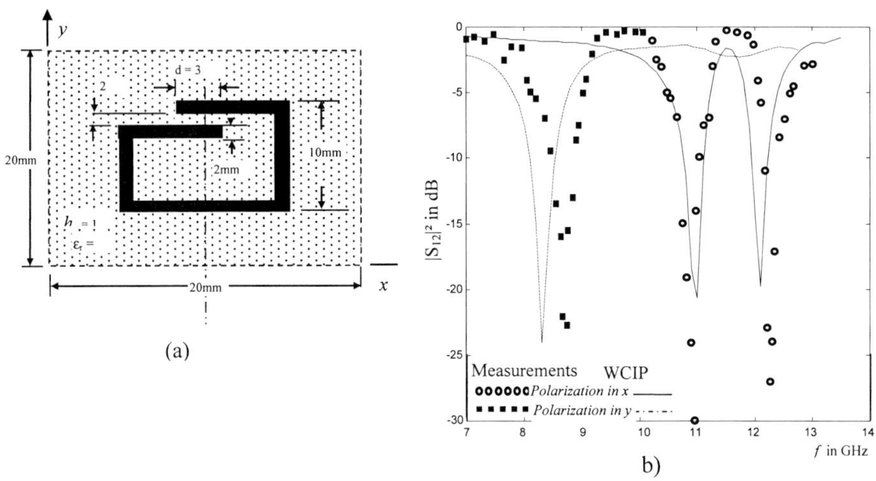

Figure 3.11. *FSS with an open quasi-square ring [TIT 09]: a) geometry of the unitary cell; b) transmission coefficient*

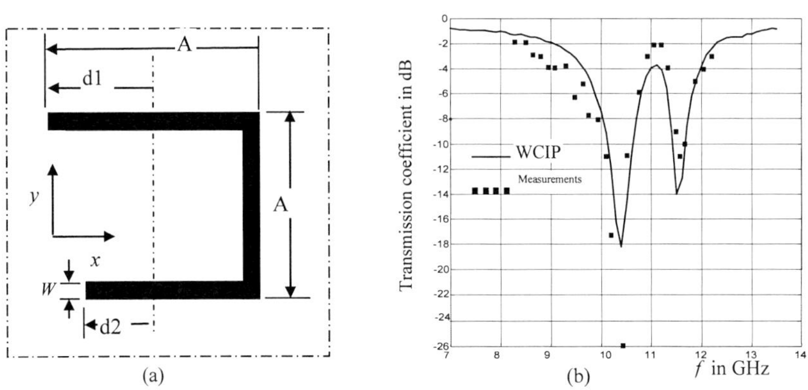

Figure 3.12. *Metallic pattern U-shaped FSS with dimensions of: A = 10 mm, d_1 = 2.5 mm, d_2 = 5 mm,* W = 2 mm, *ε_r = 4 and h = 1mm [TIT 09]: a) unitary cell; b) transmission coefficient according to the operating frequency*

3.3.4. *Double-layer FSS plating*

Double-layer FSS plating makes it possible to achieve integrated structures enabling stacked topologies, which are required both within the technology of integrated circuits [WAN 03, BAU 03] and new applications where the target antenna is accounted for with FSS structure in modeling, to take account of mutual interactions [MON 07]. In general, the addition of an interface containing a planar circuit translates as the addition of a resonance. This may also lead to an improvement in the rejected band (or band pass), provided the two resonances are sufficiently near to each other.

3.3.4.1. *Rectangular FSS rings across the two interfaces*

The double-layer plated FSS structure under consideration is that of Figure 3.4 with $\varepsilon_{r1} = \varepsilon_{r4} = 4.5$. This is a substrate of glass fibers labeled "FR-4". The thickness of both substrates is 1 mm. An air layer with a thickness of 3.3 mm separates both the interfaces Ω_1 and Ω_2. The external square dimension length of the FSS square ring at the interface Ω_1 and that of the interface Ω_2 are respectively $A_1 = 9.8$ mm and $A_2 = 8.2$ mm. The width of the metal strip within both of the FSSs is 2 mm. The transmission coefficient of this FSS is shown in Figure 3.13.

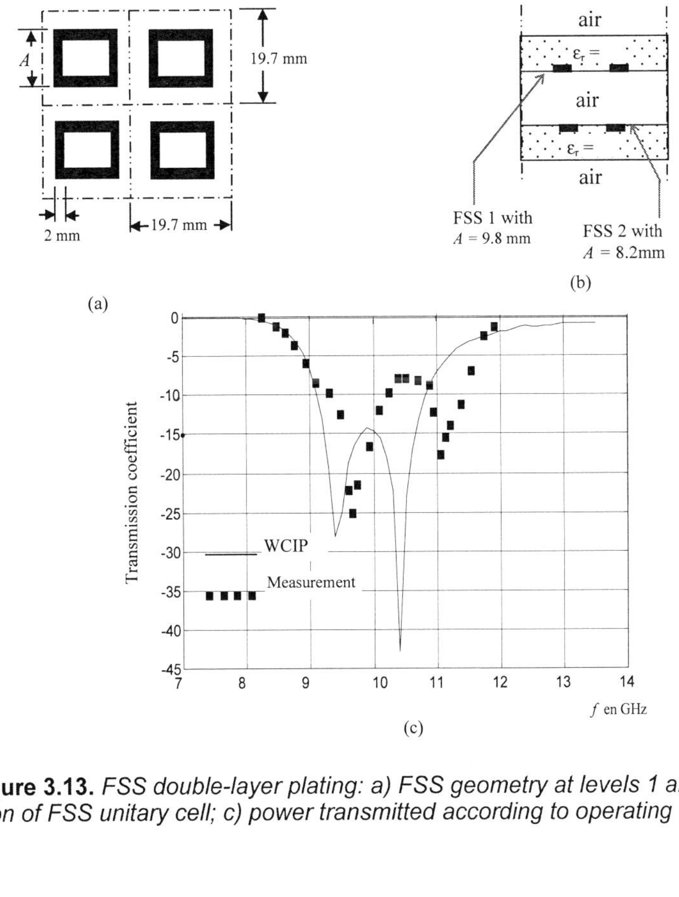

Figure 3.13. *FSS double-layer plating: a) FSS geometry at levels 1 and 2; b) section of FSS unitary cell; c) power transmitted according to operating frequency*

3.3.4.2. *Broadband FSS*

Multilayer plated FSS devices may afford a broad band of rejection, provided a judicious choice of FSS configuration is made. This type of FSS may serve as a ground plane for transmitter broadband antennas [ERD 02].

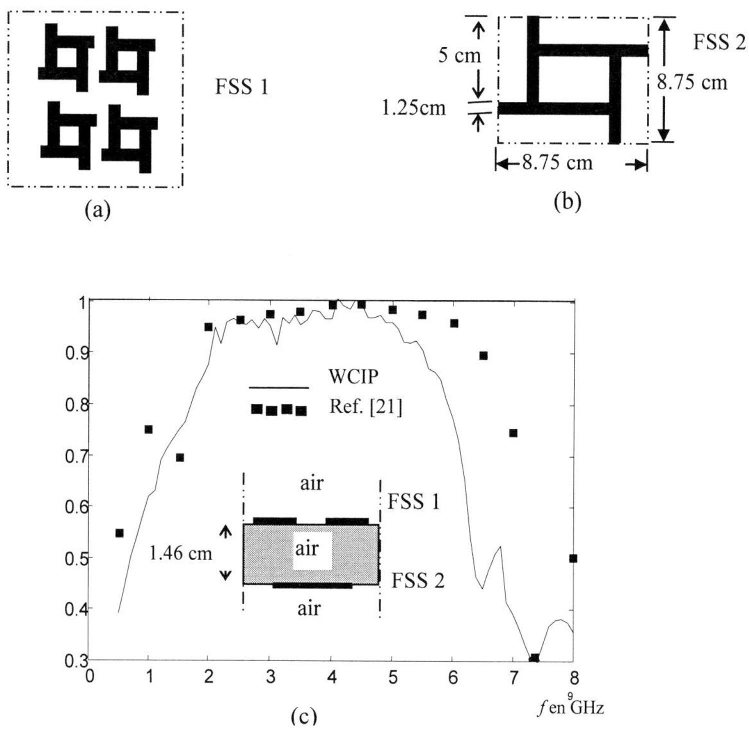

Figure 3.14. *Metallic configuration FAN type FSS [ERD 02] with double-layer plating: a) FSS 1; b) FSS 2; c) FSS Reflection coefficient*

3.3.5. *Triple-layer plating*

An increase in FSS bandwidth may be achieved by adding an additional level of plating. In Figure 3.15, identical triple-layer FSS plating is studied through using the WCIP method. The unitary FSS cell etched on the triple-layer plating FSS is shown in Figure 3.15(a). Figure 3.15(b) shows a section of the resulting FSS. Its characterization by the WCIP method is based upon the extension of the FSS double-layer plating approach to that of the triple-layer plating approach, by resorting to inter-level adjacent transition operator specified in [3.17].

Five poles can be noted within the increased bandwidth as Figure 3.15 shows, when characterizing an FSS with identical rectangular slits etched on three levels of plating. Two poles occur due to the resonances of the dielectric thickness, which separates the metal screens. The other poles result from both the resonances of the metal screen and the symmetry displayed by the structure.

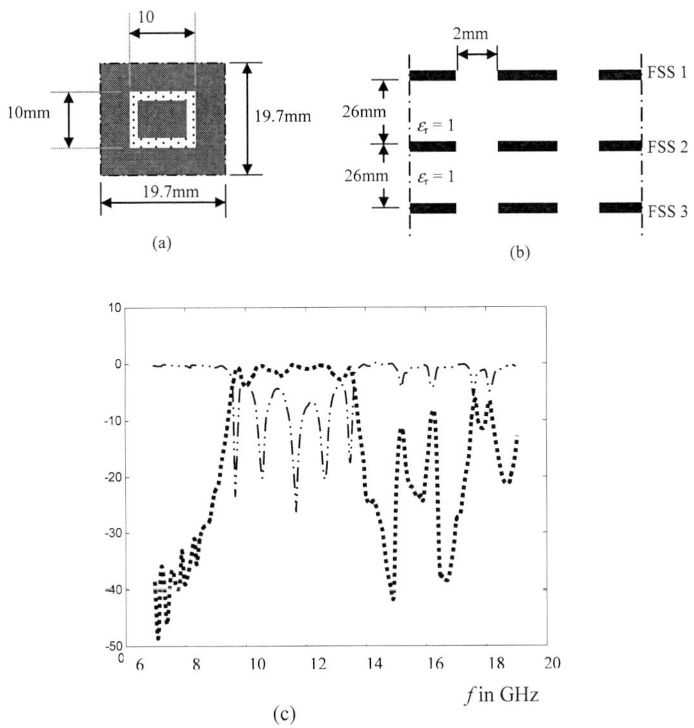

Figure 3.15. *Triple-layer plated FSS with identical square slits: a) unitary cell of a square-slit FSS; b) triple-layer plated FSS section; c) reflection coefficient with dielectric rings triple-layer plating FSS obtained by WCIP*

3.3.6. *Thick FSSs*

In reality, the thickness of the metal within FSSs is not always insignificant. As a consequence, significantly thick FSS devices have been studied for several decades [LUE 78]. Improvements in the WCIP method to analyze thick FSSs have been made. Within the thick perforated metallic screens, the screen is considered as a network of identical metallic waveguides. Within the WCIP method this thick screen introduces a level of additional plating and the problem to be resolved is that of double-layer FSS plating. However, in this case both plated layers are enclosed within a non-periodic but metal waveguide.

The thick multi-layer plated FSSs afford structures, which behave as a pseudo-elliptic filter. This is a type of filter which is significant in industry due to its straight band transition characteristics from the pass-band to the stop-band and of the number of transmission zeros that can be adjusted.

3.3.6.1. *WCIP method adapted to thick FSS devices*

Figure 3.16 shows an FSS with dielectric slits based upon a metal screen placed upon a dielectric substrate, and fed by a plane wave with a normal incidence. The virtual dotted lines in Figure 3.2 represent periodic walls, restricting the FSS unitary cell. A section of the FSS unitary cell FSS under consideration is shown in Figure 3.17. It consists of three so-called regions which are formulated as region I, region II and region III.

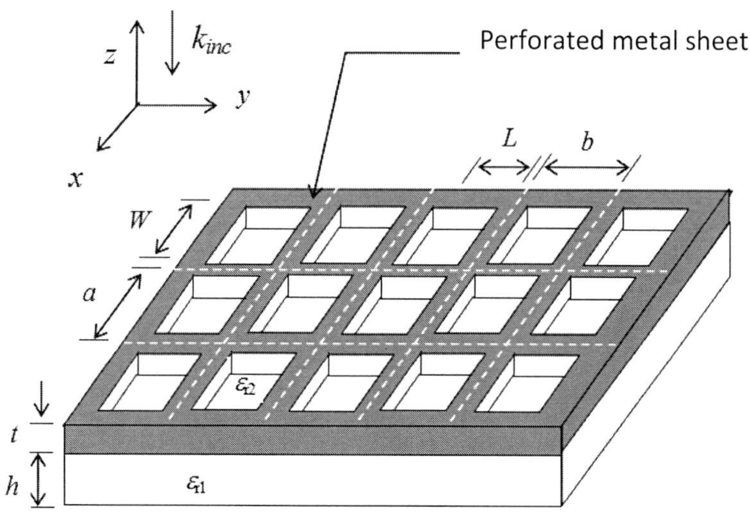

Figure 3.16. *FSS thick slit resonant FSSs*

The dielectric layer made up of air and substrate in Figure 3.17 is converted into a single WCIP layer. This is done by reducing the admittances of the TE and TM modes of the air layer to the interface separating region III from region II passing through the substrate thickness layer *h* [TIT 7a]. The resulting model shown in Figure 3.18 is the WCIP problem containing three FSS thick slit resonant media.

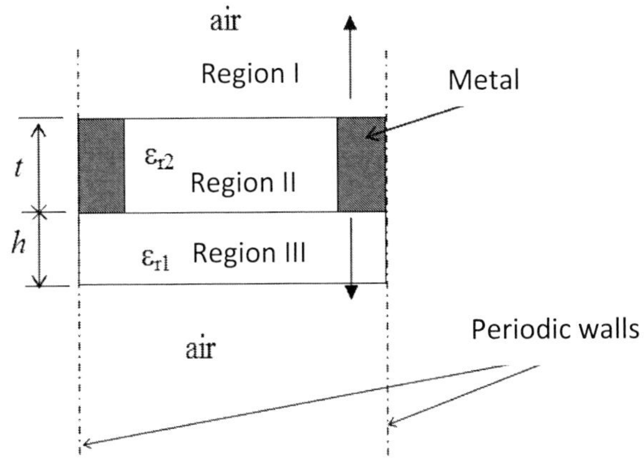

Figure 3.17. *Section of the unitary cell of the thick FSS device*

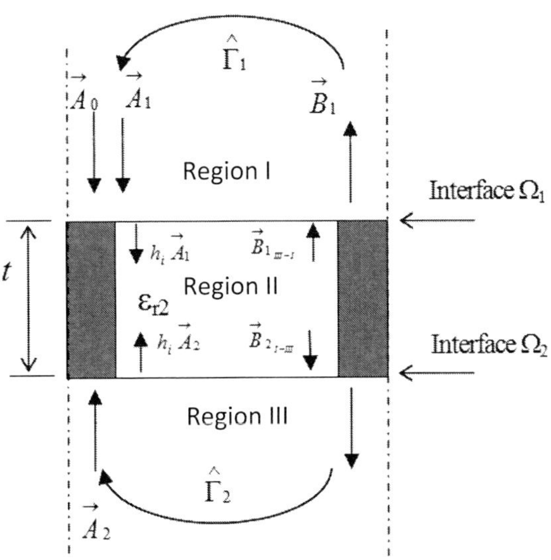

Figure 3.18. *The three-layer media WCIP issue containing resonant FSS thick slits*

Regions I and III are restricted by periodic walls. In these two regions, the conventional WCIP method, adapted to the FSS, is used as explained in detail in [TIT 07a, TIT 07b]. The wave diffracted in region I (or region III) is equal to the part of the wave diffracted from the metallic domain plus the wave coming from region III (or region I) through the rectangular-shaped metallic waveguide. This may be expressed as the equation:

$$\begin{cases} \vec{B}_1 = -h_m \vec{A}_1 + \vec{B}_{1_{III-I}} \\ \vec{B}_2 = -h_m \vec{A}_2 + \vec{B}_{2_{I-III}} \end{cases} \quad [3.18]$$

where h_m is the indicator function of the metal domain. $\vec{B}_{1_{III-I}}$ and $\vec{B}_{2_{I-III}}$ are respectively waves coming from region III and arriving in region I, and waves leaving region I to reach region III through region II. Region II is modeled by a rectangular metallic waveguide with a length t being the thickness of the perforated metallic screen. $\vec{B}_{1_{III-I}}$ and $\vec{B}_{2_{I-III}}$ are expressed as:

$$\begin{vmatrix} \vec{B}_{1\,III-1} \\ \vec{B}_{2\,I-III} \end{vmatrix} = \left[\hat{S}_T \right] \begin{vmatrix} h_i \vec{A}_1 \\ h_i \vec{A}_2 \end{vmatrix} \quad [3.19]$$

with h_i being the indicator function of the dielectric domain filling the metal waveguide, possessing a rectangular shape which models the thickness of the perforated metallic sheet.

$\hat{S}_T$ defined within the modal domain is the transition operator crossing both interfaces Ω_1 and Ω_2, determining the extent of the perforated metal sheet as shown in Figure 3.18. It is expressed by:

$$\left[\hat{S}_T^{\alpha} \right] = \begin{vmatrix} \dfrac{\left(Z_\alpha^2 - Z_{0_{II}}^2\right)\sinh\left(\gamma_{2,mn} t\right)}{\Delta^\alpha} & \dfrac{2 Z_\alpha Z_{0_{II}}}{\Delta^\alpha} \\ \dfrac{2 Z_\alpha Z_{0_2}}{\Delta^\alpha} & \dfrac{\left(Z_\alpha^2 - Z_{0_{II}}^2\right)\sinh\left(\gamma_{2,mn} t\right)}{\Delta^\alpha} \end{vmatrix} \quad [3.20a]$$

$$\Delta^\alpha = 2 Z_\alpha Z_{0_2} \cosh\left(\gamma_{2,mn}\, t\right) + \left(Z_\alpha^2 + Z_{0_2}^2\right)\sinh\left(\gamma_{2,mn}\, t\right) \quad [3.20b]$$

where α designates TE_{mn} modes and TM_{mn}, Z_α is the impedance of the mode TE_{mn} or the mode TM_{mn} of the metallic surface rectangular waveguide, which is modeled upon region II. Z_{0II} is the characteristic impedance of the dielectric filling the waveguide, limited by the two interfaces Ω_1 and Ω_2, and $\gamma_{II,mn}$ is the propagation constant of the mode TE_{mn} or the mode TM_{mn} within region II.

The generating functions of the modes TE and TM of a rectangular metallic waveguide are shown in [WAN 03] described in the following equations:

$$\begin{cases} f_{x,mn}^{TE}(x,y) = \frac{-n}{b} C_{mn} \cos\left(\frac{m\pi x}{a}\right) \sin\left(\frac{n\pi y}{b}\right) \\ f_{y,mn}^{TE}(x,y) = \frac{m}{a} C_{mn} \sin\left(\frac{m\pi x}{a}\right) \cos\left(\frac{n\pi y}{b}\right) \\ f_{x,mn}^{TM}(x,y) = \frac{m}{a} C_{mn} \cos\left(\frac{m\pi x}{a}\right) \sin\left(\frac{n\pi y}{b}\right) \\ f_{y,mn}^{TM}(x,y) = \frac{n}{b} C_{mn} \sin\left(\frac{m\pi x}{a}\right) \cos\left(\frac{n\pi y}{b}\right) \end{cases} \qquad [3.21]$$

where the following equations also apply:

$$C_{mn} = \frac{1}{\sqrt{\left(\frac{m}{a}\right)^2 + \left(\frac{n}{b}\right)^2}} \sqrt{\frac{2\tau_{mn}}{ab}} \qquad [3.22]$$

$$\tau_{mn} = \begin{cases} 2 \text{, si } (m,n) \neq (0,0) \\ 1 \text{, si } m=0 \text{ or } n=0 \end{cases}$$

Incident waves $\vec{A_1}$ and $\vec{A_2}$ across interfaces Ω_1 and Ω_2 are projected upon the entire modal base [3.21] to determine the amplitudes for modes TE and TM, giving rise to waves that reach regions I and III through metal wave guides replicating the outline of the perforated metal sheet.

To maintain the characteristic of the WCIP method, as regards reduced computational time, through the use of the exponential FFT2 algorithm, a FFT2 algorithm based upon the functions *cosi* and *sico* must be derived from the exponential form of an FFT2, being the bi-dimensional Fast Fourier Transform specified in [TIT 07a]. The functions *cosi* and *sico* are stated in:

$$\begin{cases} \text{cosi}\,(x,y) = \cos\left(\dfrac{m\pi x}{a}\right)\sin\left(\dfrac{n\pi y}{b}\right) \\ \text{sico}\,(x,y) = \sin\left(\dfrac{m\pi x}{a}\right)\cos\left(\dfrac{n\pi y}{b}\right) \end{cases} \qquad [3.23]$$

The FFT2 based upon the functions *cosi* and *sico*, which characterize the metallic rectangular waveguides, is obtained directly by doubling the structure restricted by periodic walls in the x direction and in the y direction, taking into consideration the wave direction to check the boundary conditions, as shown in Figure 3.19. The region in gray in Figure 3.19 represents the rectangular waveguide restricted by periodic walls.

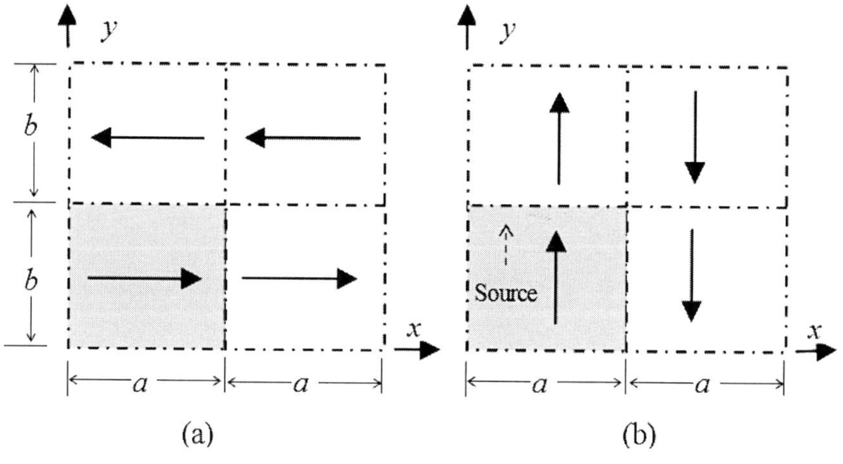

Figure 3.19. *Wave direction $h_i \vec{A}_1$ and $h_i \vec{A}_2$ in the structure when doubled a) according to the x axis direction; b) according to the y axis direction*

Incident waves across the structure when doubled in the x axis direction and in the y axis direction are inserted in [3.19] to determine the waves reaching regions I and III, and coming, respectively, from regions III and I via region II.

3.3.6.2. *Validation of the improved WCIP method*

The FSS structure in Figure 3.16 is analyzed by the method WCIP in which the perforated metal sheet has a thickness of t = 0.0175 cm and slits of dimensions W = 1.32 cm and L = 0.128 cm. The dimensions of the unitary cell are 1.78 cm x 1.78 cm. The FSS structure is fed by a normally incident plane wave and polarized in the y direction. The FSS structure to be analyzed may be used as a radome where a superior dielectric material (the so-called superstrate) may be added to insulate the antenna compartment [MON 07]. The resulting structure was made up of four dielectric layers in which the bottom layer and the top layer are both made up of air. The WCIP multilayer problem [TIT 08] should be applied to bring the TE and TM modes admittance of the air layers to the interfaces Ω_1 and Ω_2 through the dielectric slabs covering the perforated metal sheet which are the superstrate and the substrate, resorting to the model of the length of the guide. The interfaces Ω_1 and Ω_2 and the FSS unitary cell are each divided into 128 × 128 pixels and the iterative method is halted after 1,200 iterations.

Figure 3.20 shows the power transmitted and the power reflected according to the thick FSS operating frequency based uniquely upon the perforated metal sheet with a thickness t = 0.175 mm, without either a superstrate or a substrate and without a dielectric medium in the holes. The WCIP approach which is improved and adapted to thick FSS devices is validated in Figure 3.20. In filling the slits by an insulator with a dielectric constant $\varepsilon_r = 4$, a reduction in the resonance frequency shown in Figure 3.21 was observed comparative to the resonance frequency of the same FSS but without any dielectric in the slits, seen in Figure 3.20.

In Figures 3.22, 3.23 and 3.24 the power transmitted relates to the perforated metallic sheet placed upon a substrate and covered with a superstrate both of the thickness h and of the same dielectric constant ε_r. The resonance frequency is inversely proportional to the thickness h of the substrate and that of the superstrate.

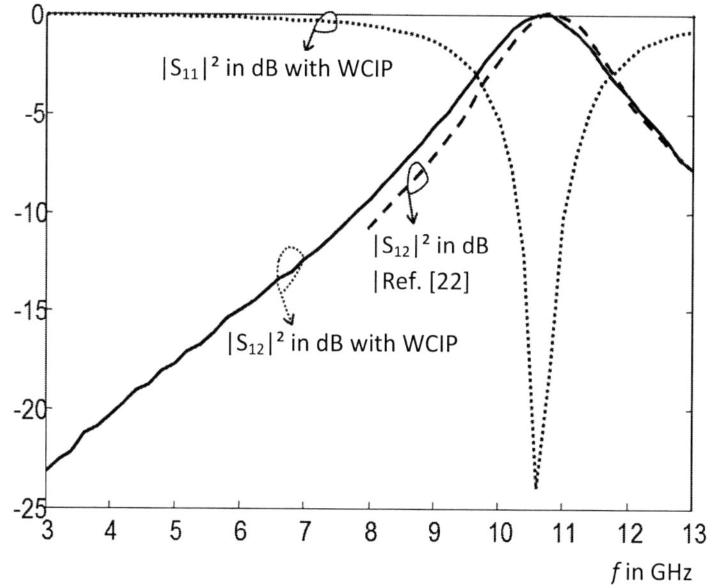

Figure 3.20. *Power both transmitted and reflected according to frequency in the case of a thick FSS with slits filled with air*

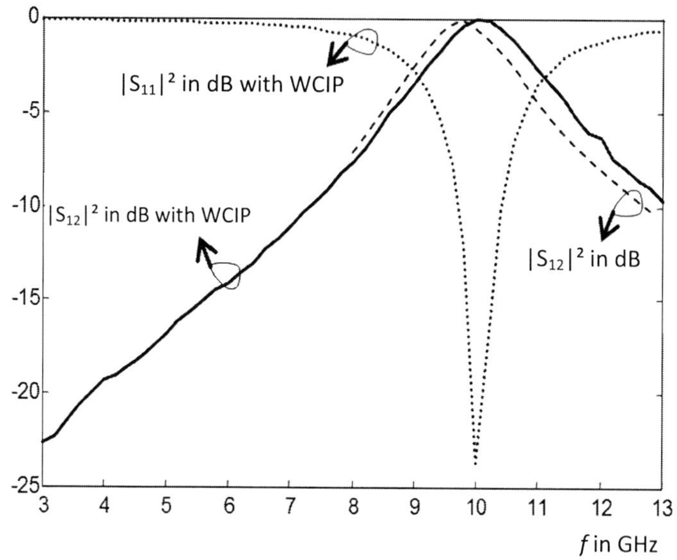

Figure 3.21. *Power outputs transmitted and reflected according to frequency in the case of a thick FSS with slits filled with a dielectric medium $\varepsilon_r = 4$*

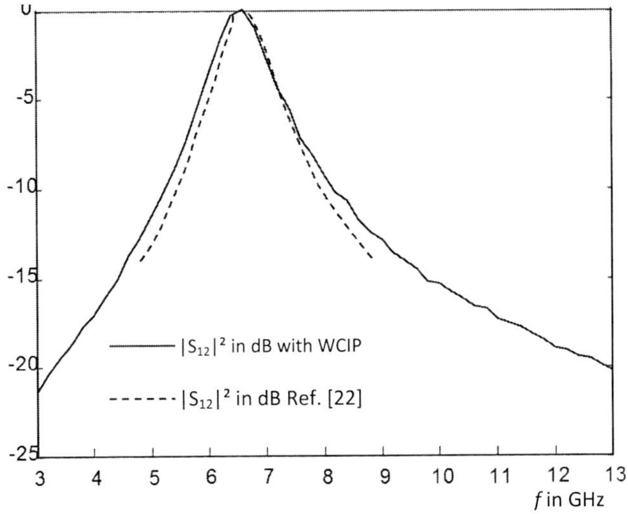

Figure 3.22. *Power transmitted through a thick metal sheet with rectangular slots sandwiched between the substrate and a superstrate both being identical and having $\varepsilon_r = 4$ and h = 0.79 mm*

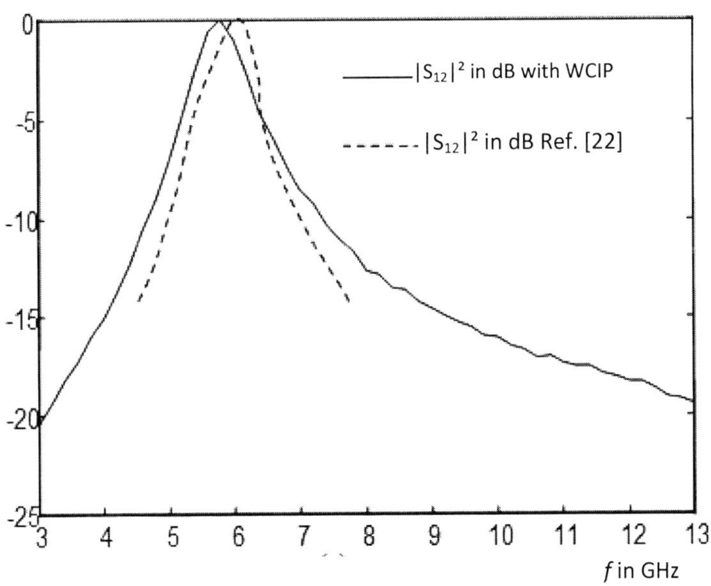

Figure 3.23. *Power transmitted through thick metal perforated sheets with rectangular slits sandwiched between a substrate and a superstrate both the same having $\varepsilon_r = 4$ and h = 1.585 mm*

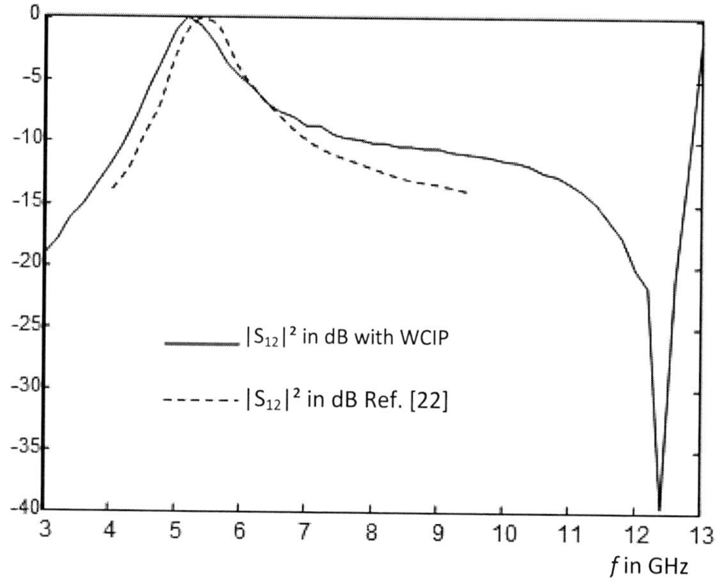

Figure 3.24. *Power transmitted through a thick metal perforated sheet with rectangular holes and sandwiched between the substrate and a superstrate both identical with ε_r = 4 and h = 3.17 mm*

The increase in the thickness h results in an increase in the effective dielectric constant of regions I and III with the WCIP triple media issue in Figure 3.18. Consequently, the resonance frequency diminishes as it is inversely proportional to the effective dielectric constant. This property of the material may be applied for an extensive adjustment of the FSS structure. The results of improved WCIP method agree with the simulation results in [MON 07].

To study the effect of the thickness of the perforated metal sheet upon diffraction parameters of the resonant FSS, the power of the non-dielectric FSS, both transmitted and reflected, which is displayed for different values for the sheet thickness t, are shown in Figure 3.25. The power transmitted by the conventional WCIP method and the improved and adapted WCIP method for thick FSSs, for t = 0 coincides to validate the approach of the thick FSS for the WCIP method since the thin FSS which has the value t = 0 is only one particular case of the thick FSS if t = 0. In Figure 3.25 it may also be observed that the resonance frequency of the thick FSS with rectangular holes is directly proportional to the thickness of the perforated metal sheet within the thick FSS, which allows the fine-tuning of the FSS resonant frequency.

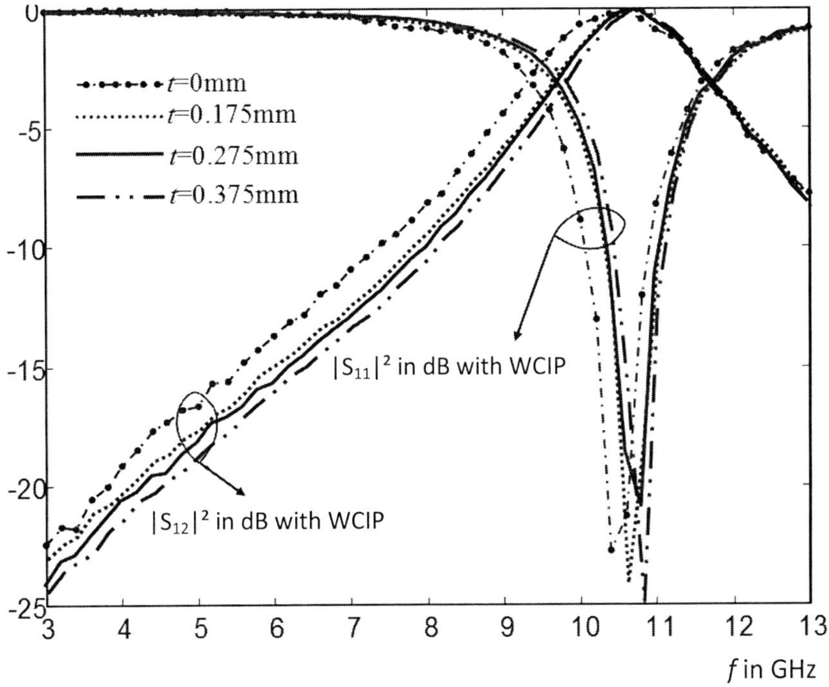

Figure 3.25. *Powers both transmitted and reflected according to the frequency of a thick FSS in the absence of an insulator in the holes for different thickness values of the perforated metal sheet*

Due to the maximal transmission of the power around the resonance for the case of the thick FSS without the dielectric of Figure 3.25, the hole of the unitary cell of the FSS may be represented schematically by a capacitance C in parallel with an inductance L. The capacitance is due to the fringe effects of the rectangular cavity, modeling the metal waveguide of the intermediate region II. Since the majority of modes within the rectangular hole are evanescent, the fields at the interfaces Ω_1 and Ω_2, which come out of regions I and III, increase when the length of the hole t decreases.

Consequently the capacitance C increases and the resonant frequency of the thick FSS reduces.

The power which is both transmitted and reflected in Figure 3.26 relates to the thick FSS structure with rectangular holes, filled with an insulator with a relative dielectric constant $\varepsilon_r = 4$ for different thickness values for the perforated metal sheet. The perforated metal sheet filled with dielectric material is sandwiched between the air layers. The resonant frequency of the resulting FSS structure diminishes when the thickness of the metal sheet t increases. Hence the possibility of broadening the precision control band of the resonant frequency is present when in the simultaneous presence of both this behavior and that shown in Figure 3.25.

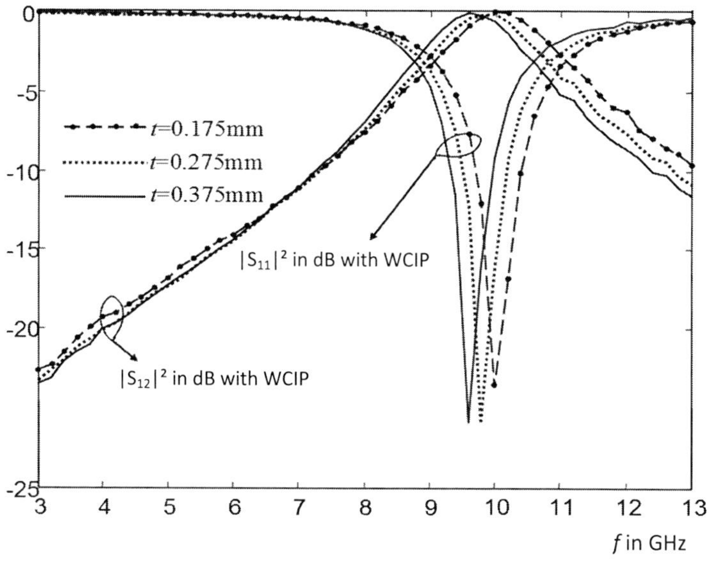

Figure 3.26. *Powers which are both transmitted and reflected according to the frequency of a thick FSS with an insulator in the holes with $\varepsilon_r = 4$ for different thickness values for the perforated metal sheet*

In this case the hole consists of a single propagating mode. As a consequence, an increase in capacitance resulting from the presence of dielectric material covering the slits (rectangular-shaped holes) leads to a reduction in the resonant frequency. The same behavior is observed in Figures 3.20 and 3.21, when the insulator is inserted in the holes of the thick FSS sheet.

3.4. Anisotropic FSS

The anisotropic dielectric layers are characterized by less loss comparative to isotropic dielectrics. For the case study relating to the uniaxial anisotropic dielectrics, the approach of the isotropic layer equivalent to the dielectric anisotropic layer is presented in [HOR 80], using a quasi-static approach. A metal strip etched on an anisotropic uniaxial substrate placed upon a metal ground plane is shown in Figure 3.27. The uniaxial dielectric constant of the anisotropic substrate is given in [3.24]:

$$[\varepsilon] = \begin{bmatrix} \varepsilon_{xx} & 0 & 0 \\ 0 & \varepsilon_{yy} & 0 \\ 0 & 0 & \varepsilon_{zz} \end{bmatrix} \quad [3.24]$$

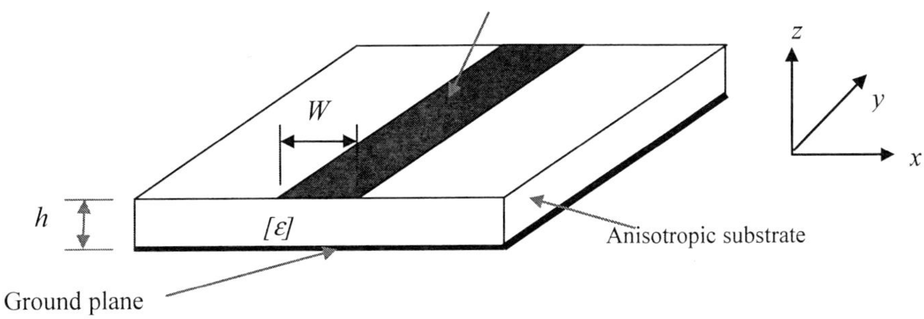

Figure 3.27. *Transmission line on an anisotropic substrate with a system of coordinates*

In the case where $\varepsilon_{xx} = \varepsilon_{yy}$ the equivalent dielectric layer may have an equivalent thickness h_{eq} and a dielectric constant ε_{eq} given by [HOR 80], and the following equations apply:

$$h_{eq} = h\sqrt{\frac{\varepsilon_{xx}}{\varepsilon_{zz}}} \quad [3.25]$$

$$\varepsilon_{eq} = \sqrt{\varepsilon_{xx}\varepsilon_{zz}} \quad [3.26]$$

The iterative method is used to characterize an FSS rectangular patch etched on two dielectric anisotropic layers, by adopting the equivalent layer approach presented in [3.25] and [3.26]. In addition, simulation results are given in Figure 3.28.

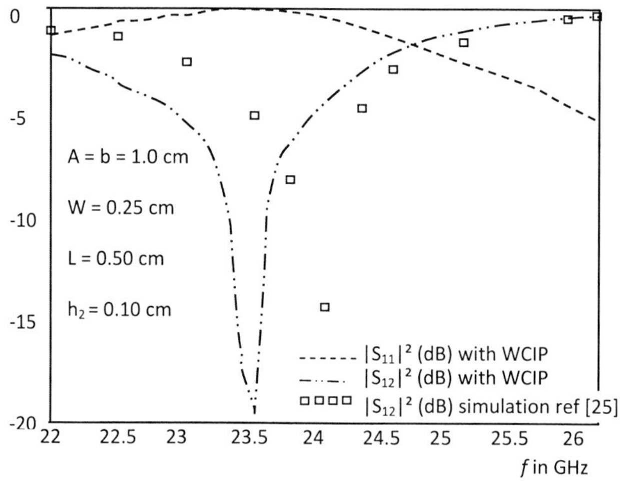

Figure 3.28. *Powers reflected and transmitted according to the operating frequency of a rectangular FSS patch etched on an air-like layer sandwiched between the patch and a dielectric anisotropic layer [TIT 07a, TIT 13]*

3.5. Measurement system

The data acquisition system applied for these measurements consists of a microwave network analyzer (of the N5230A Agilent type) which has the ability to increase to a frequency of 18 GHz and possesses two 12 dB amplification cone-shaped antenna, as shown in Figure 3.29. Measurements have been carried out upon FSS sheets of configurations of 10 × 10 unitary cells. The FSSs are made using PCB "Printed Circuit Board" technology with copper. FSS patterns are etched on the substrate of glass fibers known as "FR-4".

The difference between these measurement results and those of the WCIP is due to the fact that the substrate dielectric constant used is not precisely known; the latter may also be subject to variations when the FSS structure encompasses the various implementation stages through the use of PCB technology.

3.6. Conclusion

The WCIP method is applied to various types of FSS planar structures as follows:

1) the FSS with dielectric circular rings charged with passive and active elements for an adjustable selectivity;

2) the FSS with two metal rectangular-shaped rings, and the FSS with a U-shaped metal pattern for multiple-band applications;

3) the FSS open quasi-square ring for multiple-band applications having two perpendicular polarizations;

4) the FSS with double-layer triple-layer plating for multiple-band applications and applications which necessitate large bandwidths and/or very little volume for miniaturization purposes encountered in integrated circuits.

The results obtained with the WCIP method agree both with the measurement results and those of the works in the field.

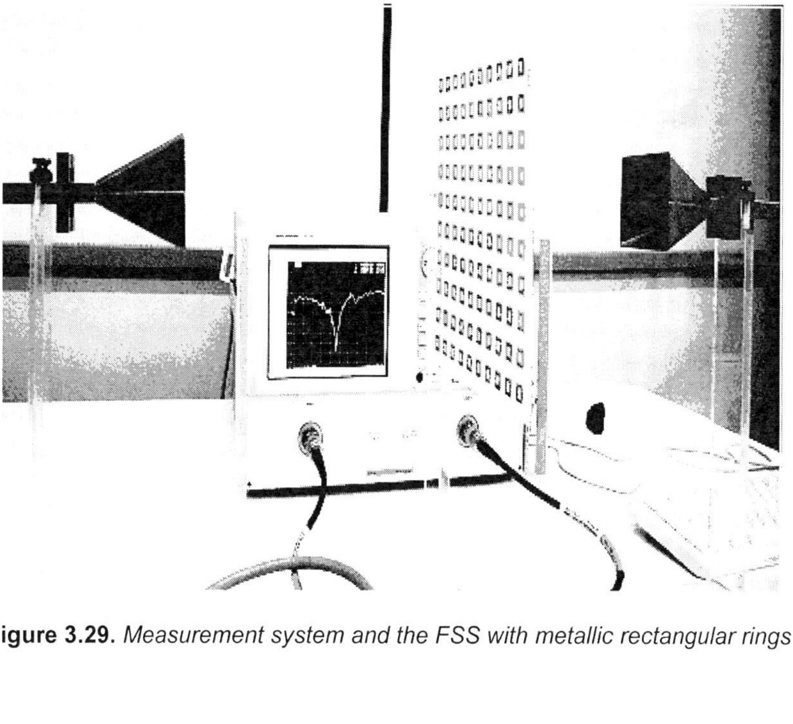

Figure 3.29. *Measurement system and the FSS with metallic rectangular rings*

Inhomogeneous layers, which are seen in integrated circuits, are due to the presence of DT "deep-tranch" and/or GR "guard ring" devices used to ensure insulation between the various circuit ports to ensure their anticipated function and to prevent undesired elements coupling. In addition, inhomogeneous layers are present in FSSs in Tera-Hertz applications, when the patterns are made of dielectrics, which are obviously of a given thickness. An integrated model with the WCIP method for thin inhomogeneous layers is being correctly tested.

The WCIP method might be extended to studying multi-layer plating FSS integrated with antennas and antenna arrays, such as arrays of metallic rectangular waveguides, to reduce the volume of the resulting structure if both are independently designed. Furthermore, the interaction between the two is inherently taken into account by the WCIP method. Such structures are necessary to prevent interferences when many communication systems are placed near each other in addition to the possibility of improving the antennas gain.

The WCIP method may also be extended to so-called 3-D (three-dimensional) problems, by using current volume density instead of tangential magnetic fields. This approach will clearly be applied to a dielectric or metal via holes-based problems encountered in substrate integrated waveguide circuits (SIWs).

3.7. Acknowledgments

The authors wish to thank Professor Alfrêdo Gomes Neto of the Telecommunications–Electromagnetism group, the Centre Fédéral de l'Education Technologique de Paraiba, João Pessoa, Brésil (the Telecommunications and Electromagnetism group at Paraiba's Federal Centre for Technological Education) for all of his measurements used as well as for supplying both the FSS and the measurement system pictures.

4

WCIP Applied to Substrate Integrated Circuits: Substrate Integrated Waveguide (SIW) and Substrate Integrated Non-Radiative Dielectic (SINRD) Circuits

4.1. Introduction

The number of electronic circuits embedded in electrical devices that we use in our daily life is constantly increasing. Integration, miniaturization and reduction of the costs of circuit production, while maintaining an acceptable level of actual device performance, both constitute technological and design challenges. Substrate Integrated Circuits (SIC) arose from the combination of two technologies [HIR 98]: planar technology which, while low-cost, involves significant losses, and so-called waveguide technology, which is high-cost but involves small losses. This technological combination corresponds to the operational performance (filter, coupler, transmission line and other circuit components) of waveguide technology within a substrate.

We will now consider the example of metallic waveguides. Metallic plates present on both sides of the substrate serve as a framework for the waveguide, and the metallic vias connected to these two plates, enable performance of the expected function, provided they are correctly

---

Chapter written by Nathalie RAVEU and Ahmad ISMAIL ALHZZOURY.

positioned. In 1994, the first Substrate Integrated Waveguide (SIW) appeared [SHI 94] which was followed in 1995, by the Low Temperature Co-fired Ceramic (LTCC) technological design [PIL 95]. Numerous applications have resulted in the design of antennas [YAN 04, CHE 08, WU 08], filters [CHE 08, WU 08, HAO 05], couplers [CAS 02, DES 02, WU 02], duplexers [HAO 05] and other related equipment. Owing to its integration capabilities, active functions are also integrated, such as oscillators [WU 03 CAS 03], mixers [HON 06] and amplifiers [JIN 08, WEN 08].

Let us now take the example of fiber-optic technology. By adding via-holes within the substrate, the effective dielectric constant is reduced. Indeed, it is possible to create, by correctly adding via-holes, various zones with different effective dielectric properties within the substrate and thus modify the electric field propagation. The Non-Radiative Dielectric (NRD) waveguide was demonstrated in 1981 by Yoneyama and Nishida [YON 81, NIS 81]. It was made up of a dielectric substrate inserted between two metallic plates. The integration within a substrate [CAS 04a, WU 04b, CAS 04b, WU 04b] only occurred from 2004 onwards, with the advent of Substrate Integrated Non-Radiative Dielectric (SINRD) structures.

The simulation of these circuits is also a challenge, as the number of vias (whether metalized or not) increases, particularly in cases where they cannot be taken into account through homogenization techniques. These circuits have been studied using different numerical methods: the Finite-Difference Time-Domain (FDTD) [SIM 04], the Method of Moments (MoM), [YAN 00], the Finite Elements Method (FEM) [ZHO 96] and more recently with a semi-analytical method based upon Green's functions [MAC 05, TAM 05, ARN 08, AME 08]. The WCIP formulation [ZAI 10] was modified so as to be adapted to these circuits (see section 4.2). The performance across various SIW and SINRD circuits is then tested across several structures (see sections 4.3 and 4.4). Finally, to conclude the chapter, future development paths are suggested in section 4.5.

4.2. Formulation of WCIP for SIC circuits

Amongst the SIW and SINRD circuits produced, several have a "regular" distribution of the vias, for ease of manufacture, in particular, but, also as a secondary issue, in the interest of homogenization [ISM 11, ISM 12, ISM 14]. The Wave Concept Iterative Procedure (WCIP) method does not claim to characterize all possible types of the SIC circuits but only those for which vias (whether or not metallic) are spread over a regular grid with each cube having the widths $d_x \times d_y$ following the x and y directions. The entire structure is thus able to be represented across a grid with the dimensions $D_x \times D_y$, and being made up of elementary cells, as shown in Figure 4.1.

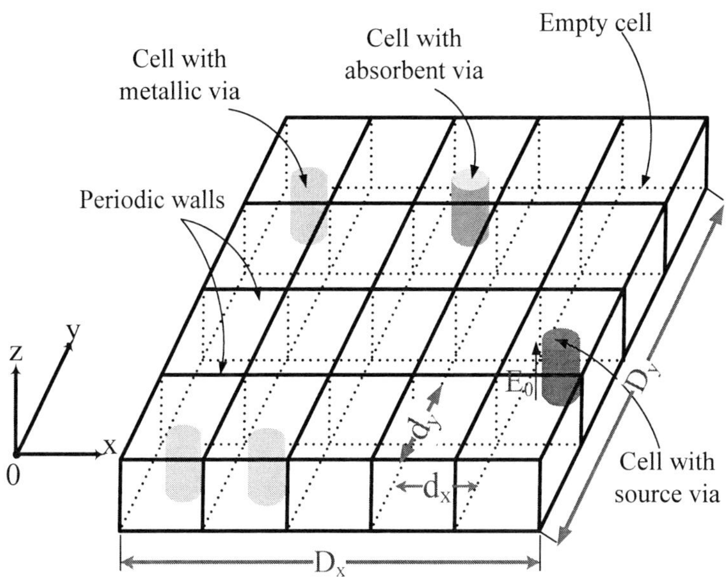

Figure 4.1. *Configuration of the SIC circuits across a regular grid*

The circuit under study should be restricted on its four lateral sides (as the top and the bottom are plated), so as to define the computational domain. Generally they are assigned a so-called "fictive" boundary condition, in the sense that it does not correspond to a real physical boundary. This process is not problematic as these limits can be pushed back far beyond the field of study of antennas. These do not impact upon the results for non-radiative circuits. The external boundaries of the circuit may take three forms: metallic, magnetic or periodic. The circuit is then divided into cells of equal dimensions. The waves are therefore defined at cell level, by a B incident wave across this via and an A wave, which is diffracted by this via. The unknowns across the entire circuit are grouped into vectors and marked as **A** and **B**. Their dimensions consist of the total number of cells within the structure.

Each cell within this grid, in fact, corresponds to a pixel within the conventional WCIP. At its center, the cell consists of a via of radius r_{a}. The method does not therefore allow the examination of the SIC circuits displaying vias with different radii. The boundary conditions upon the cell are, in fact, those upon this via. Such conditions may be of several types: metallic, dielectric having the same permittivity as the substrate, or dielectric with a different permittivity from the substrate and source. A diffraction operator $\hat{S}$ takes into account the conditions across all of the cells of the circuit with the equation [4.1].

$$\mathbf{A} = \hat{S}\mathbf{B} \tag{4.1}$$

On account of boundary conditions upon the lateral sides of the circuit, waves may be decomposed on a modal base, with a generating function F_{pq}, ascertaining the external boundaries without being concerned as to the presence of vias. The existence of vias, which are assumed to be periodic, enables us to define at each cell a specific modal base which corresponds to the cell, from the generating function $f_{pq,mn}$. From this, we are then able to calculate the disruption across the modal base of the entire structure. A diffraction operator $\widehat{\Gamma}$ takes into account the modal conditions across the entire grid of circuit cells considered by [4.2].

$$\mathbf{B} = \widehat{\Gamma}\,\mathbf{A} \tag{4.2}$$

Lumped sources $\mathbf{A_0}$ are taken into account in the equation within the spatial domain [4.1], while modal sources $\mathbf{B_0}$ are taken into account in the equation within the modal domain. In the examples considered in this chapter only lumped sources are used. The equation to be solved is ultimately that in [4.3]. It is resolved by the Generalized Minimal Residual Method (GMRES) [SAA 86]:

$$\left(\mathbf{I} - \hat{S}\widehat{\Gamma}\right)\mathbf{A} = \mathbf{A_0} \tag{4.3}$$

4.2.1. *The definition of* $\hat{S}$

Boundary conditions across the elementary cell, positioned on the grid in (i,j), depend upon the type of via present at its center:

– for an empty cell $S(i,j) = 1$

– for a metallic cell $S(i,j) = -1$

– for a cell exhibiting relative dielectric permittivity ε_g,when the relative permittivity of the substrate is ε_r, $S(i,j) = (\varepsilon_g - j(\varepsilon_g - \varepsilon_r))/(\varepsilon_g + j(\varepsilon_g - \varepsilon_r))$. It is noted that if $\varepsilon_g = \varepsilon_r$ then $S(i,j) = 1$ (case of an empty cell)

– for an absorbent cell $S(i,j) = 0$

– for a source cell $S(i,j) = (Z_{0s}-Z_0)/(Z_{0s}+Z_0)$ with Z_{0s} the source internal impedance.

4.2.2. *the definition of* $\hat{\Gamma}$

The modal scattering operator [4.4] is defined using the modes F_{pq}, which correspond to a circuit having the dimensions $(D_x \times D_y)$. The number of modes is taken as being equal to the number of cells within the structure:

$$\hat{\Gamma} = \sum_{pq} |F_{pq}\rangle \Gamma_{pq} \langle F_{pq}|, \qquad [4.4]$$

with the equations:

$$F_{pq} = \frac{1}{\sqrt{D_x D_y}} e^{j\frac{2\pi p}{D_x}x} e^{j\frac{2\pi q}{D_y}y}, \qquad [4.5]$$

$$\Gamma_{pq} = \frac{Z_{pq}-Z_o}{Z_{pq}+Z_o}. \qquad [4.6]$$

The mode impedance is defined according to [4.7]. A convergence study in both n and m is made for each given pq pair using the equation:

$$Z_{pq} = \sum_{m,n} \frac{j\omega\mu_0 |\langle H|f_{pq,mn}\rangle|^2}{k_0^2\varepsilon_r - \alpha_{p,m}^2 - \beta_{q,n}^2}, \qquad [4.7]$$

with k_0 being the wavenumber in free space; μ_0 the substrate permeability; ω the pulsation; H an indicator function for position, shape and size of the via in the cell; $f_{pq,mn}$ the generating function of the modal base of the elementary cell with dimensions ($d_x \times d_y$) with a boundary consisting of periodic walls [4.8] using the following equations:

$$f_{pq,mn} = \frac{1}{\sqrt{d_x d_y}} e^{j\alpha_{p,m}x} e^{j\beta_{q,n}y} \qquad [4.8]$$

$$\alpha_{p,m} = \frac{2\pi p}{D_x} + \frac{2\pi m}{d_x}$$

$$\beta_{q,n} = \frac{2\pi q}{D_y} + \frac{2\pi n}{d_y}$$

where $(p,q,n,m) \in Z^4$ are the propagation constants, respectively in x and in y.

4.3. Results for SIW circuits

Several SIW structures are characterized by WCIP using the formulation explained in section 4.2. The results obtained are compared with a commercial code known as Ansys–High Frequency Structural Simulator (HFSS) based around the Finite Elements Method, in terms of both accuracy and computation time.

4.3.1. *Waveguides*

We will now consider the SIW waveguide as shown in Figure 4.2. Metallic vias serve as electric walls. The effective width of the SIW waveguide is denoted w_{eff}, the physical distance between vias (also called the center distance), from one side to the other of the waveguide is denoted by w.

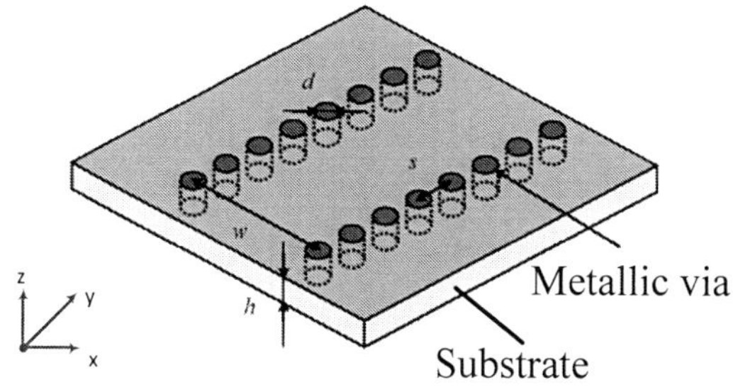

Figure 4.2. *The SIW waveguide with a fundamental mode TE_{10}*

The width w_{eff} depends on the frequency and the vias configuration. It may be extracted from the waveguide transmission coefficient. For this particular SIW waveguide, the fundamental propagation mode is a TE_{10} mode. The wavelength λ [4.9] of the waveguide may be assessed by measuring the phase of the transmission coefficient S_{21} between the two sources, placed at its extremities with the formula in [4.10].

$$\lambda = \frac{c}{\sqrt{f_0^2 \varepsilon_r - f_c^2}} \qquad [4.9]$$

with f_0 being the frequency, c the light celerity and f_c the cut-off frequency of the TE_{10} mode $f_c = c/2(w_{eff})$,

$$\lambda = \frac{-2\pi L}{\angle S_{21}} \qquad [4.10]$$

Metallic vias have a center distance of 2 mm ($= s = d_x = d_y$) and their diameter is $d = 1.25$ mm. The width of the waveguide w is 40 mm and its length L is 68 mm. The substrate Arlon AD255A (TM) is used with a relative permittivity of 2.55, a loss tangent of 0.0015 and a thickness of 1.524 mm. According to [ISI 13], the effective width is given by the formula in [4.11].

$$w_{eff} = w - \frac{(d)^2}{0.95p} \qquad [4.11]$$

In this case, a w_{eff} of 39.8757 mm is obtained with equation [4.11]. This width may also be determined at each frequency numerically with the assistance of the equations [4.9] and [4.10], from which we may then extract the equation in [4.12] as follows:

$$w_{eff} = \frac{\lambda_0}{2\sqrt{\varepsilon_r - \left(\frac{\lambda_0 \angle S_{21}}{2\pi L}\right)^2}} \qquad [4.12]$$

with λ_0 being the vacuum wavelength.

It may be seen in Figure 4.3 that the cut-off value obtained when using the WCIP tends toward the theoretical value of 39.8757 mm, but that the effective width also depends on the frequency. This only remains a single mode structure across a narrow band above 2.5 GHz. In addition, above a particular frequency the distance between metallic vias no longer serves as a metallic wall and allows for the escape of the electromagnetic field.

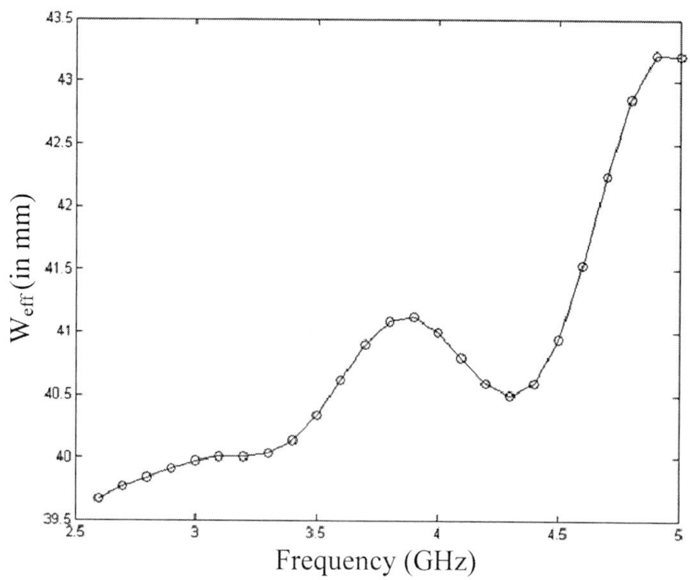

Figure 4.3. *Effective width of the waveguide for the TE_{10} mode according to frequency*

4.3.2. *Bandpass filter*

The design of the SIW bandpass filter is inspired by the metallic waveguide design, in which both walls and sides that delimit the apertures are substituted by metallic vias. The distance between the vias should be sufficiently small to make the actual conditions similar. It is then necessary to reduce the number of metallic vias delimiting the apertures while maintaining the filter performances. The initial input sources (generally waveports) are then substituted by coaxial cables driven into the substrate with a central core connected to one of the metallic plates and the ground connected to the other metallic plate. The filter is finally bounded by electric walls, in the place of the initial sources (waveports), which enable matching but also improve the quality coefficient. The second order filter is shown in Figure 4.4. The device has an operating frequency of 3.8 GHz, a relative bandpass width of 15% and a ripple of 0.05 dB.

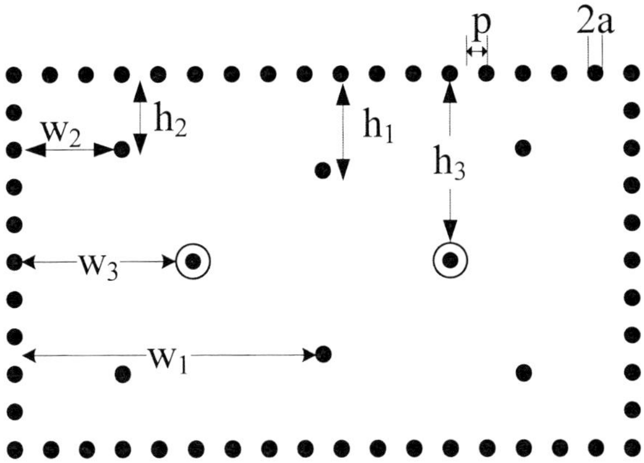

Figure 4.4. *Second-order SIW bandpass filter dimensions, w_1 = 34 mm, h_1 = 10 mm, w_2 = 12 mm, h_2 = 8 mm, w_3 = h_3 = 20 mm, p = 4 mm, a = 0.625 mm*

The results obtained are compared with the measurement and the HFSS simulations as shown in Figure 4.5. The measurements and both simulations results (WCIP-HFSS) are in good agreement. The simulation time is 1.86 s/frequency for the WCIP compared with 37 s/frequency for HFSS Central Processing Unit (CPU): Intel Core 2 Duo E6550 at 2.33 GHz, with a RAM of 4 Gb).

In each of the structures studied here, the WCIP offers similar results in terms of accuracy compared with the Finite Elements Method. Nevertheless, with the number of metallic via being low, the computation times prove to be comparable. The WCIP's ease of use and reduced computation time has made possible sensitivity studies upon the shape of the metallic vias, the number of vias to place on the edge of the filter and on the sides that delimit the apertures [ISM 11, ISM 12].

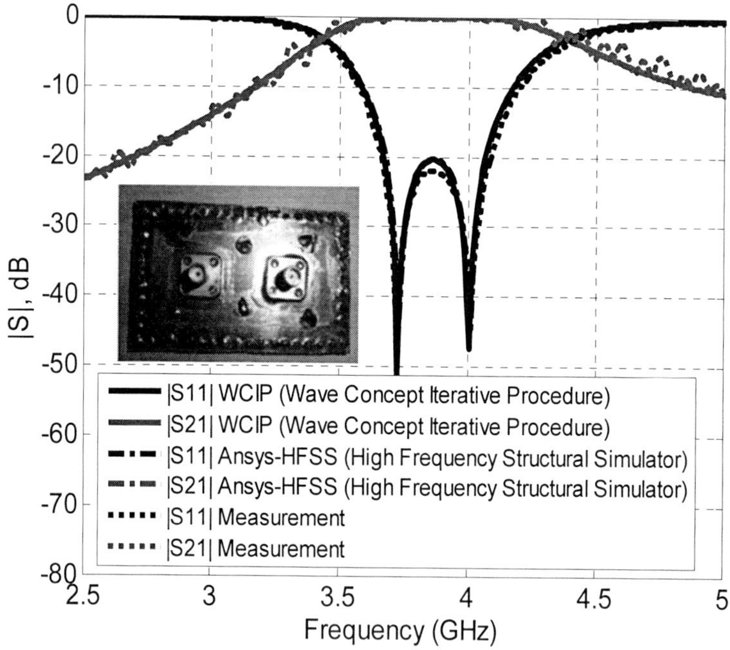

Figure 4.5. *Measurements and simulations (WCIP-HFSS) results of the S parameters of the second-order filter shown in Figure 4.4*

4.4. Results for the SINRD circuits

Several SINRD structures are characterized by the WCIP [ISM 14] with the formulation which is explained in section 4.2. The results obtained are compared with the same commercial code Ansys–HFSS (the commercial finite element method solver system for electromagnetic structures). To be able to carry out the simulation with the Ansys–HFSS, the zones, where the via-holes are drilled, are substituted by an effective dielectric permittivity zone (ε_{eff}) which is equivalent. Indeed the number of via-holes is too important to simply allow a simulation, if this homogenization condition is not included in the Ansys–HFSS method.

This homogenization condition is obtained by the simulation of the TEM mode in the structure shown in Figure 4.6.

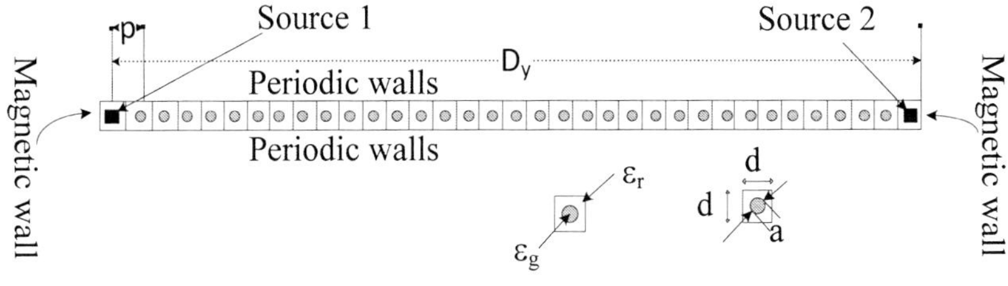

Figure 4.6. *Waveguide of fundamental mode TEM through the SIRND zone*

The effective dielectric constant is directly deduced from the propagation constant according to the equation in [4.13] for $L = 60$ mm.

$$\beta = k_0 \sqrt{\varepsilon_{eff}} = \frac{-\angle S_{21}}{L} \qquad [4.13]$$

The effective relative permittivity of the dielectric is represented along the frequency in Figure 4.7 for a distance between the via-holes (center distance) of 2 mm. The substrate is still Arlon AD255A.

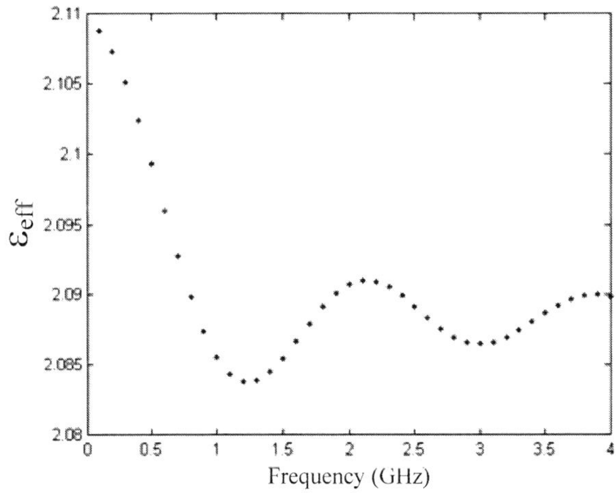

Figure 4.7. *Determination of effective relative permittivity*

The simulation results are in good agreement with the approximation which is made in [ISI 13] which states the figure of 2.0745 for this cell. The use of the TEM waveguide nevertheless seems better suited as it enables the determination of the effective relative permittivity value for all useful frequencies while in [ISI 13] the value is only obtained at the cut-off of the waveguide for the fundamental mode TE_{10}.

4.4.1. *Waveguides*

The electromagnetic field is generally confined in high permittivity dielectric zones. This is the principle of fiber-optics. The Longitudinal Section Electric (LSE) and Longitudinal Section Magnetic (LSM) waveguides benefit from this principle so as to create new modes of propagation. In this section, the LSE and LSM waveguides are fabricated by drilling through the edges of the substrate as shown in Figure 4.8(a). Taking into account the number of holes, the media is replaced by a homogeneous media of effective relative permittivity. This is shown in Figure 4.8(b) in respect to HFSS simulation. The cut-off frequencies of the inherent modes are compared for several waveguide core widths noted ℓ. The results are shown in Table 4.1.

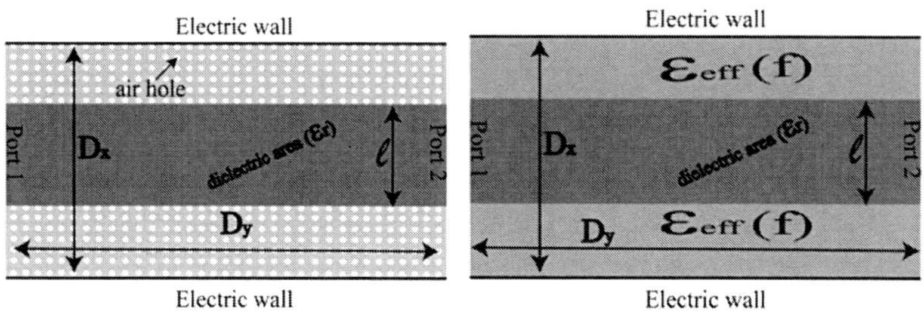

Figure 4.8. *LSE-LSM waveguides; a) WCIP; b) equivalent guide for HFSS; D_x = 40mm; D_y = 68 mm*

ℓ/ Dx f_c (GHz)	0.25	0.35	0.45	0.55	0.65	0.75
HFSS	2.37	2.33	2.29	2.27	2.25	2.24
WCIP	2.369	2.328	2.298	2.276	2.26	2.25
Relative error (in %)	0.04	0.085	0.34	0.26	0.44	0.44

Table 4.1. *Cut-off frequencies and relative error between the HFSS and WCIP for different LSE-LSM waveguide core widths ℓ*

The computation time is of 0.34 s/frequency for the WCIP against 1.36 s/frequency for HFSS for a width ratio of l/Dx = 0.25 (using the CPU: Intel Core 2 Duo E6550 at 2.33 GHz, with a RAM of 4 Gb).

4.4.2. *Bandpass filter*

The second order bandpass filter specifications include a center frequency of 3.8 GHz, a ripple of 0.05 dB and a relative bandpass width of 15%. The filter obtained displays two resonators and an inverter. It is shown in Figure 4.9.

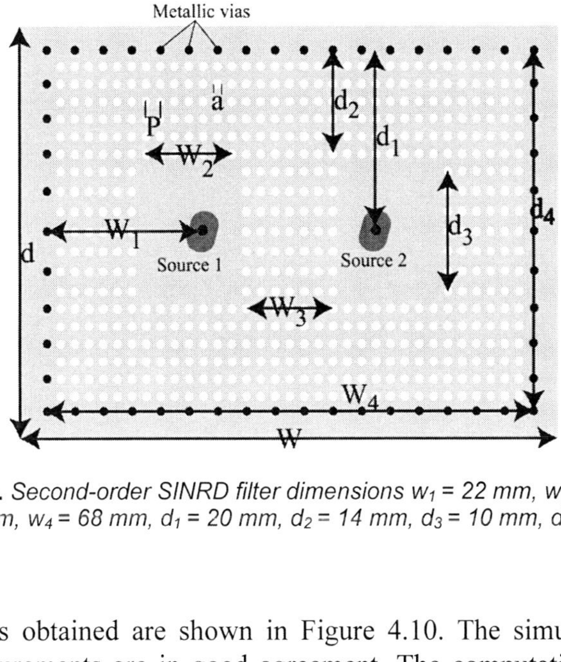

Figure 4.9. *Second-order SINRD filter dimensions w_1 = 22 mm, w_2 = 12 mm, w_3 = 14 mm, w_4 = 68 mm, d_1 = 20 mm, d_2 = 14 mm, d_3 = 10 mm, d_4 = 40 mm*

The results obtained are shown in Figure 4.10. The simulation results and the measurements are in good agreement. The computational time for the WCIP is 4.3 s/frequency, while it is 6.22 s/frequency for the HFSS (using the CPU: Intel Core 2 Duo E6550 at 2.33 GHz and the RAM of 4 Gb), which includes, in addition, homogenized substrate conditions for via-hole drilled zones, whereas the WCIP simulates each via-hole independently.

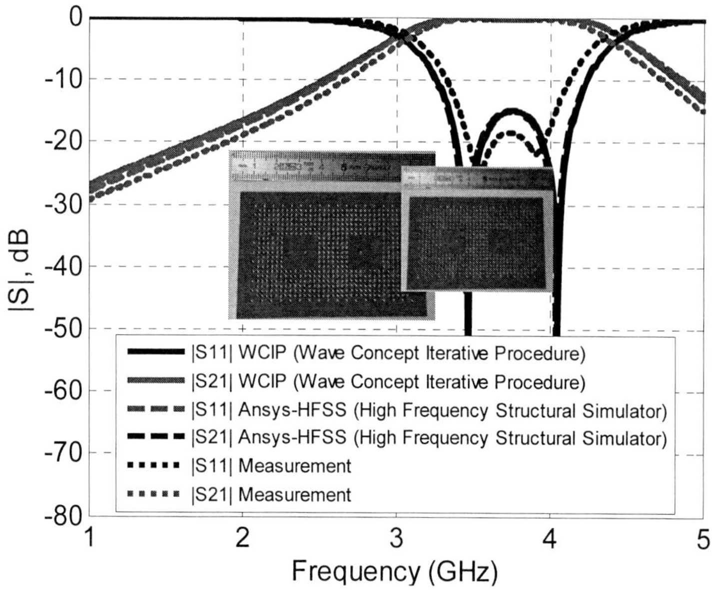

Figure 4.10. *Measurement and simulation (both HFSS and WCIP) results of the S parameters of the second-order filter shown in Figure 4.9*

In this section, we listed the advantages of WCIP in terms of accuracy and computation time compared with the Finite Element Method, which requires the homogenization of the zone with via-holes to enable simulation.

4.5. Conclusion

The possibility of combining the metallic vias and the dielectric vias concepts within the design of passive functions without being limited by the number of via-holes is one of the major assets of WCIP, which thus competes with both the Method of Moments and the Finite Elements Method. The limitations of this method lie in the hypotheses stated initially: that the vias should all have the same radii and they should be positioned across a regular grid layout. Nevertheless, this method, while not taking account of homogenization conditions, enables us to accurately describe a large number of circuits.

Its performance speed, linked to optimization codes, also enables a reduction in the number of via-holes during manufacture, as was tested with the SINRD filter shown in Figure 4.9. We thus obtain the filter as seen in Figure 4.11 (see inset *r*), the majority of holes were removed in the periphery of the circuit, being the zone where the electric field is weak within the filter bandpass. The measurements and simulations are in very good agreement (please see Figure 4.11).

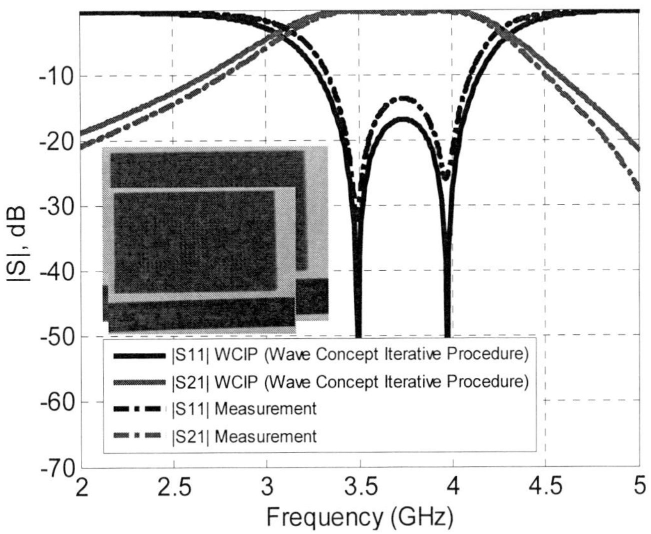

Figure 4.11. *Simulation (WCIP) and measurement results of the S parameters of the non-homogeneous SINRD filter*

The use of WCIP for large-scale SIW and SINRD circuits might enable the testing of both the limitations and operational performance of this method. The design of a circuit with a non-homogeneous distribution of via-holes (while remaining upon the initial regular grid) also enables enhanced value of the equipment, when compared with other digital apparatus in existence.

5

WCIP Convergence

5.1. Introduction

For more than 15 years, the Wave Concept Iterative Procedure (WCIP) has turned out to be a promising method for characterizing multilayer circuits, demonstrating multi-layer plating (see [WAN 05, AKA 01, BAU 01]. The WCIP convergence has been carried out in the study of passive circuits as its operators are both bounded and their actual values are less than one (see [BAU 99, GON 99]). Nevertheless, convergence may be slow to take place, which makes the method uncompetitive with the existing computation codes. Initially, the WCIP is determined by a fixed-point method (see [CRU 08, RAY 08]) by directly applying the equations of which it consists. It is unfortunately probably not the best way to solve the problem. As its operators are diagonal but defined across different bases (both spatial and modal), base changes are necessary between operators. The overall operator is therefore never formulated. Thus the solution is approached by using successive vector matrix products.

Furthermore, several parameters enter into the wave definition, or boundary conditions. These boundary conditions are fixed according to physical considerations, such as the wave impedance reference, source

---

Chapter written by Nathalie RAVEU.

impedance, and perfect boundary conditions across the metal. These parameters may be adjusted to reduce the computation time.

In this chapter, several WCIP convergence acceleration techniques are both suggested and tested upon a single example. The results produced by the software Ansys–HFSS for this example will be taken as a reference point in precise terms. Computation times, as well as the number of iterations, are set out at each new improvement stage.

5.2. Summary of WCIP

WCIP was developed to solve Helmholtz equations for printed circuit boards on a homogeneous substrate. The circuit is enclosed in a box with metallic, magnetic or periodic walls (or any combination of these) as shown in Figure 5.1. The equations to solve for this configuration are stated in equation [5.1]:

$$\begin{cases} \left[\Delta + k_0^2 \varepsilon_r\right] u = 0 \\ u_{\delta\Gamma_1} = 0 \quad \text{or} \quad \dfrac{\partial u_{\delta\Gamma_1}}{\partial n} = 0 \\ u_{\delta\Gamma_2} = 0 \quad \text{or} \quad \dfrac{\partial u_{\delta\Gamma_2}}{\partial n} = 0 \\ u_{\delta\Gamma} = 0 \quad \text{or} \quad \dfrac{\partial u_{\delta\Gamma}}{\partial n} = 0 \\ u_{\delta S} = 0 \quad \text{and} \quad \dfrac{\partial u_{\delta S}}{\partial n} = 0 \end{cases} \qquad [5.1]$$

where u describes the electric or magnetic field; ε_r the relative permittivity of the given media; k_0 the number of waves within the vacuum and n is the conventional normal vector to the surface under consideration. The boundary conditions depend upon the circuit equation to solve.

The operators are defined in relation to incident waves **A** and reflected waves **B** [5.2] in place of both transverse electric fields $\mathbf{E_T}$ and transverse magnetic fields $\mathbf{H_T}$. Such operators ensure method convergence as their spectral radius is less than one. Both pairs (**A**,**B**) and ($\mathbf{E_T}$, $\mathbf{H_T}$) confirm the system [5.1]. Z_0 is generally the impedance of the vacuum, $n_{\delta S}$ is the normal vector to the δS interface through the following formulae:

$$\begin{cases} A = \dfrac{1}{2\sqrt{Z_0}}\left(E_T + Z_0 H_T \wedge n_{\delta S}\right) \\ B = \dfrac{1}{2\sqrt{Z_0}}\left(E_T - Z_0 H_T \wedge n_{\delta S}\right) \end{cases} \qquad [5.2]$$

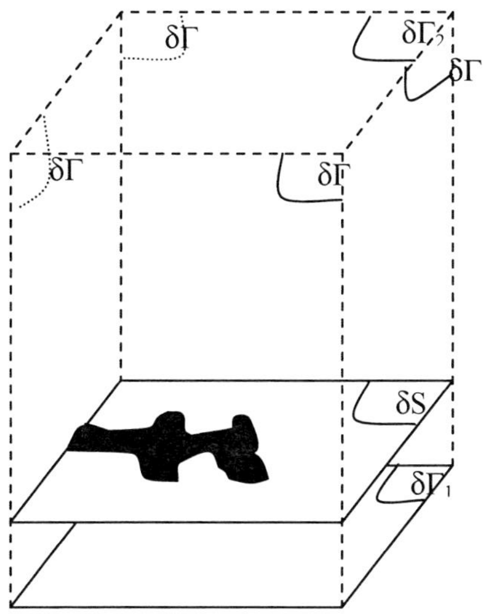

Figure 5.1. *Configuration of the circuit where fields are solved using the WCIP*

The model is thus applied to each δ*S* interface and we distinguish two types of operator: one which takes into account the homogeneous media around the interface and one which accounts for the boundary conditions at the interface.

5.2.1. *Representation of homogeneous materials around the interface*

As the circuit is enclosed within a box of finite dimensions, the fields below and above this interface may be described using a modal base. The expression of fields depends upon the conditions around the boundary $\delta\Gamma$ (whether electric, magnetic or periodic). The homogeneous material is described by a spectral operator which takes into account boundary conditions on $\delta\Gamma_1$ and $\delta\Gamma_2$ (according to [BAU 99, GON 99]). At such boundaries, electric, magnetic or open conditions may be considered. Finally around the δS interface, the system [5.1] is partially described by two operators $\hat{\Gamma}_1$ and $\hat{\Gamma}_2$ respectively for the materials 1 and 2:

$$\hat{\Gamma}_i = \sum_{n,m} \sum_{\alpha=TE,TM} \left| f_{nm}^{\ \alpha} \right\rangle \Gamma_{nm}^{\ i,\alpha} \left\langle f_{nm}^{\ \alpha} \right| \qquad [5.3]$$

where $f_{nm}^{\ \alpha}$ is the function of the modal base of the mode α (= TE or TM) of n and m; $\Gamma_{nm}^{\ i,\alpha}$ is the reflection coefficient of the mode α of the order n and m, with $\Gamma_{nm}^{\ i,\alpha} = \left(1 - Z_0 Y_{nm}^{\ i,\alpha}\right) / \left(1 + Z_0 Y_{nm}^{\ i,\alpha}\right)$ and $Y_{nm}^{\ i,\alpha}$ the mode admittance α of the order n and m.

5.2.2. *Description of boundary conditions at the interface*

The interface δS is meshed into a regular pattern in such a way as to simply transfer from the spatial base to the modal base through Fourier transforms and their inverse, linked by their transition matrices. Boundary conditions are modified by each mesh component (whether metallic, dielectric or indeed the source) according to the circuit type being characterized owing to the spatial operator $\hat{S}$. Across the source, the waves having the same polarization are totally absorbed.

5.2.3. *System to solve*

The system [5.1] is therefore also expressed by the system [5.4] which is applied to incident and reflected waves defined in [5.2]:

$$\begin{cases} \mathrm{A} = \hat{S}\mathrm{B} + \mathrm{A}_0 \\ \mathrm{B} = \hat{\Gamma}\mathrm{A} \end{cases} \qquad [5.4]$$

with A_0 the lumped source within the circuit $\hat{\Gamma} = \begin{pmatrix} \hat{\Gamma}_1 & 0 \\ 0 & \hat{\Gamma}_2 \end{pmatrix}$.

This system is generally solved by using Richardson's method; the number of meshes is identical to the number of modes and iterations are initiated from the term source. An average is calculated upon the 10 last *S* parameters values along the iterative process and serves as criteria for convergence; it helps to more rapidly reach convergence.

5.3. Improvement of WCIP by mathematical techniques

First, we must recall some of the rules that must be applied to the WCIP to obtain accurate results. Result accuracy is directly related to the mesh size, where the order of convergence of the method is 1 and the method should therefore confirm given conditions:

– the mesh elements should have side lengths which are either inferior or equal to $\lambda/100$ at each frequency of the simulation;

– both lines and slots should be made up of a minimum of five meshes in both width and length.

Generally, the mesh is established according to the highest frequency of a given circuit. Nevertheless, as operators in the modal domain should be amended for each frequency, the mesh is also redrawn at each frequency, according to the previously stated criteria.

The example dealt with is a suspended microstrip line (positioned in the air) with a stub placed in parallel as shown in Figure 5.2. The rear ground plane is placed at a distance of 0.5 mm. As the reflection (S11) and transmission (S21) parameters are direct functions of the length of the stub, the accuracy of the resonance frequencies (through S11 or S21) attest from the method accuracy. The frequency range of the study is 2.5–6 GHz with 40 frequencies points. With the mesh criteria previously stated, the interface δS is meshed with step in x and in y of identical length 0.125 mm, which does not change within the frequency band of the study. The problem has 192,000 unknowns. Tests are carried out to achieve convergence criteria of 0.01 upon the relative residual method computed across the waves of the entire interface δS. The results are obtained using an Intel Xeon 3.2 GHz, RAM 16Gb, 12-core processor machine.

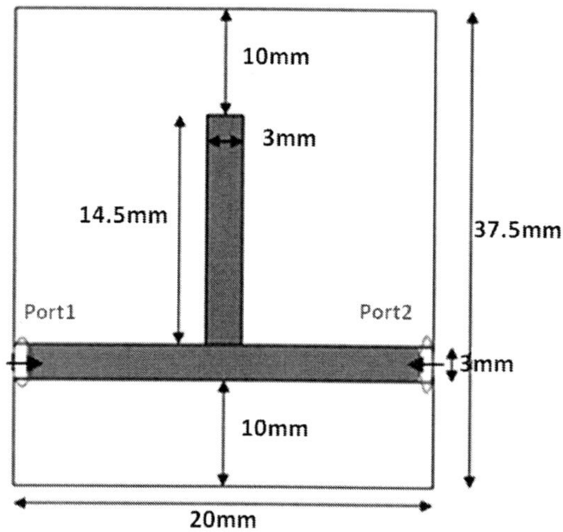

Figure 5.2. *Tested circuit*

5.3.1. *Number of modes/number of meshes*

Usually, the number of meshes and the number of modes are taken to be identical. As modal coefficients are obtained by a Fourier transform, the accuracy of the higher order modes is questionable. A linear wave interpolation within the spatial domain is carried out before each Fourier transform to better approach these coefficients, thus "numerically" doubling the number of meshes. Correspondingly, the number of modes is also doubled. Modal accuracy is also improved. These modes are maintained to complete the inverse Fourier transform. The waves are then recovered solely across the initial meshes.

An interpolation is therefore added to the iteration, which increases the computational time of each iteration but allows for a significant reduction in the number of iterations. For the required convergence criteria, the solution using Richardson's method does not allow convergence whatever the frequency (the maximum number of iterations is set to 20,000). The interpolation is therefore adopted throughout the rest of this study.

5.3.2. *GMRES/Richardson*

Although less than one [BAU 99 GON 99], the convergence radius associated with Richardson's procedure is often close to one. Hence, convergence is very slow. To reduce this convergence time, a Generalized Minimal Residual Method (GMRES) [SAA 86, SCH 86] solution is adopted. In the rest of this study, a restart level of 50 and a maximum number of 9,000 iterations for GMRES is considered. Figures 5.3(a) and (b) show convergence curves for a relative residual current obtained, respectively, using the Richardson and GMRES procedures, at a frequency of 2.5 GHz. The convergence criteria (0.01) is indicated by a threshold marked in red upon the curves. Note that on the GMRES convergence curve, the specific points around the restart at 50.

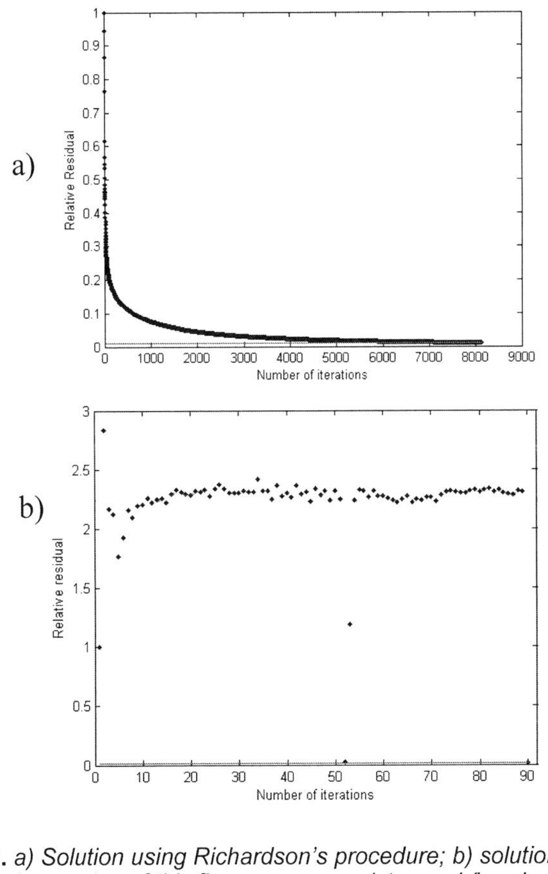

Figure 5.3. *a) Solution using Richardson's procedure; b) solution using GMRES. For a color version of this figure, see www.iste.co.uk/baudrand/waves.zip*

In [RAV 10, GIR 10, BAU 10], the solution using Richardson's procedure occurs using 101,581 iterations (or 2,540 iterations/frequency), with a computation time of 13 h and 52 min (or 21 min/frequency). The solution using the GMRES procedure takes place in 3,316 iterations (or 82 iterations/frequency), with a computation time of 1 h 20 min (or 2 min/frequency). In using GMRES, computation time is divided by 10 and the number of iterations is divided by 30. The GMRES solution has therefore been adopted in the rest of this study.

5.3.3. *Selecting the initial value*

Usually, the initial value of the iterative process corresponds to the source term at each frequency. Several tests, involving changing this initial value, have been performed, by taking a null value or the previous frequency value if the mesh is identical. The results obtained are shown in Table 5.1.

	Initial value, the source	*Initial value, the null value*	*Initial value, the previous frequency value*
Number of iterations	3,752	3,832	1,507
Average number of iterations/frequency	94	96	38
Computation time	1 hour 25 minutes 10 seconds	1 hour 28 minutes 15 seconds	30 minutes 37 seconds
Average computation time/frequency	2 minutes 7 seconds	2 minutes 13 seconds	46 seconds

Table 5.1. *Choice of the initial value*

In choosing, as the initial value, the result obtained from the preceding frequency, computational time-saving is reduced by one-third. In addition, the number of iterations is reduced by around 60% compared with the standard case, where we take the source as the initial value. The other choices for initial values are less high-performance. Remember that the frequency step in this study is around 90 MHz. The computation time-saving is less important if the frequency step increases as seen in [RAV 11, GIR 11].

If you compare the number of iterations at each frequency in Figure 5.4, taking the source as the initial value, we note that the number of iterations varies very little in frequency, having an average of around 94 iterations/frequency. On the other hand, if one takes as an initial value the solution calculated at the previous frequency, the number of iterations fluctuates in frequency. Taking the result at the previous frequency as the initial value is highly efficient at lower frequencies but loses its efficiency at higher frequencies. The sudden increase around 5 GHz also corresponds to the resonance of the circuit being studied.

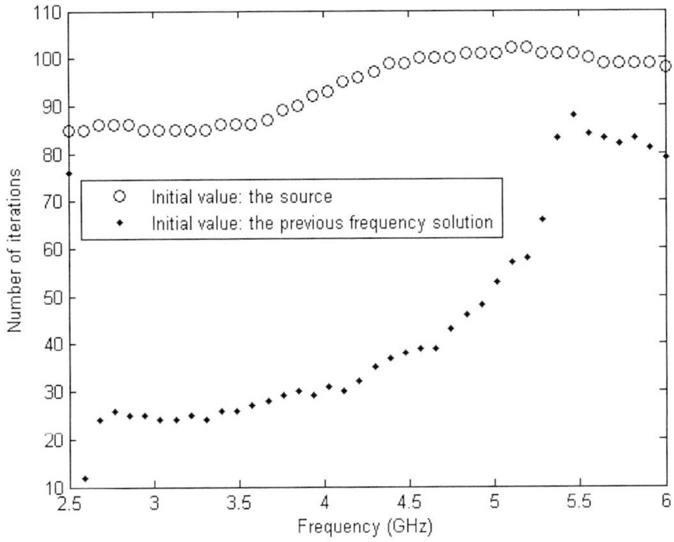

Figure 5.4. *Comparison of the number of iterations between two initial conditions at each frequency*

The use of the previous frequency result as the initial value allows us to start with a solution which is close to expected; the field distribution differs very little from one frequency to another. However as they are only approximate values, it also leads to an increase in the error as the frequency increases. A combination of both approaches is suggested; this involves taking the previous frequency as the initial value, except when the number of iterations exceeds 40. In that case the null value is again used. The results are shown in Figure 5.5. We obtain a total of 1,455 iterations in place of 1,507, a slight improvement. Within the frequency band, this initial null condition is used around 4 GHz, which allows us to maintain an average, for the highest frequencies, of around 30 iterations/frequency.

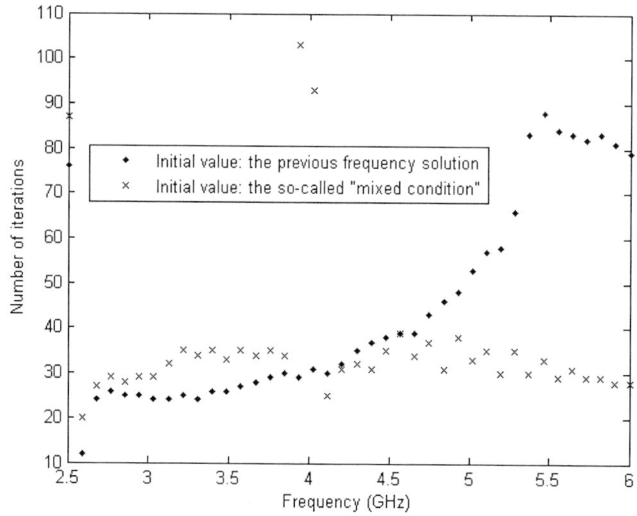

Figure 5.5. *Comparison of the number of iterations with so-called mixed conditions*

5.4. Improvement of WCIP by physical considerations

5.4.1. *Simplification of waves at the interface*

In the method shown in section 5.2, the waves are specified in a disjointed way in the materials around the interface materials δS. If we no longer consider electric and magnetic fields as within each given material but the total electromagnetic field across the interface, the operators are defined differently.

The spatial operator is only modified across dielectric meshes. If a dielectric mesh is positioned at the point (p,q) of the grid, the boundary condition is then $\mathrm{A}(p,q) = \mathrm{B}(p,q)$.

The spectral operator is simplified in [5.5].

$$\hat{\Gamma} = \sum_{n,m} \sum_{\alpha=TE,TM} \left| f_{nm}^{\ \alpha} \right\rangle \Gamma_{nm}^{\ \alpha} \left\langle f_{nm}^{\ \alpha} \right| \qquad [5.5]$$

where $f_{nm}^{\ \alpha}$ is the function of the modal base for the mode α (= TE or TM) of the order n and m; $\Gamma_{nm}^{\ \alpha}$ is the reflection coefficient of the mode α of the order n and m, $\Gamma_{nm}^{\alpha} = \left[1 - Z_0\left(Y_{nm}^{1,\alpha} + Y_{nm}^{2,\alpha}\right)\right] / \left[1 + Z_0\left(Y_{nm}^{1,\alpha} + Y_{nm}^{2,\alpha}\right)\right]$ with the $Y_{nm}^{i,\alpha}$ mode admittance α being of the order n and m of the material i.

The system to solve is still expressed by equation [5.4]. However, the number of unknowns with the same mesh is reduced to 96,000. The solution for this new formulation which we will call "simplified WCIP" occurs over 1,795 iterations (or 44 iterations/frequency). We do not therefore observe an improvement in the computation time. We even note a loss of accuracy in the results, to which Figure 5.6 attests. This formulation has thus not been maintained subsequently throughout the study.

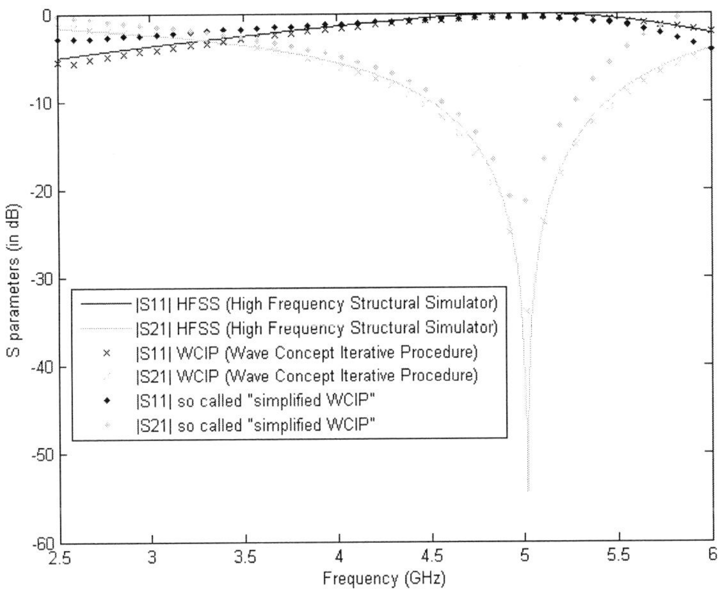

Figure 5.6. *Comparison of S parameters obtained using HFSS and the WCIP with two formulations*

5.4.2. *Choice of reference impedance*

The impedance which enters into the wave definition [5.2] and therefore the modal operator of the WCIP is usually chosen as being the impedance of the vacuum ($Z_0 = 120\pi\Omega = 377\Omega$). There is nevertheless a degree of freedom with regards to the method which can potentially influence convergence (see [RAV 09, BAU 09]). Several Z_0 values are used within the WCIP formulated in the second section. The results obtained are shown in Figure 5.7. We note that the minimum value is reached around this vacuum impedance. This particular choice is therefore retained in the rest of the study.

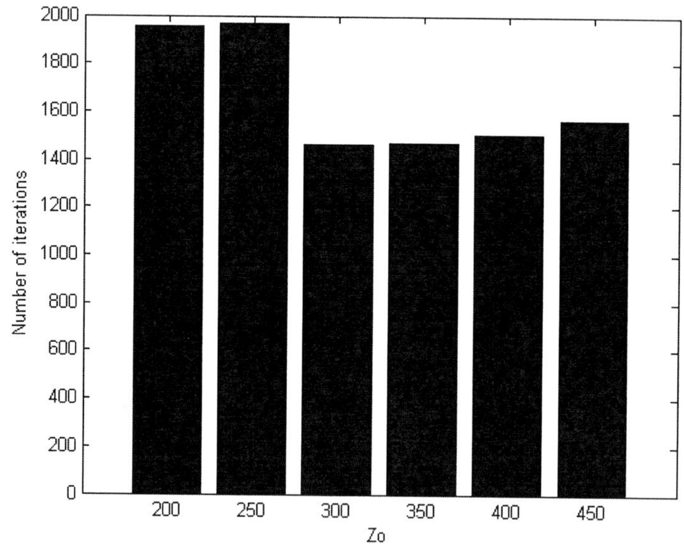

Figure 5.7. *Choice of Z_0*

5.4.3. *Boundary conditions on the metallic mesh*

The metal considered in this example is a perfect metal. Boundary conditions on the conductor is thus A = –B. The introduction of losses within the metal, even very small ones (of the order A = –0.999B), allows for further reduction in the number of iterations without changing the result obtained by both the reflection coefficient and the transmission coefficient. Indeed, the number of iteration in the case of small losses from the metal is 1,444 iterations, compared with 1,507 iterations in the case of perfect metal conditions. Nevertheless this only constitutes a small reduction (5%) compared with that achieved with other techniques.

5.5. Conclusions

Numerous techniques have been implemented to reduce the computation time of the WCIP. In normal conditions, taking convergence criteria across the relative residual leads to a very bad convergence, sometimes not converging at all. The GMRES solution, having as an initial condition the preceding calculated frequency solution, makes it possible to systematically reach convergence with a reasonable calculation time (at best 46 s/frequency). Both points constitute the best assets to reduce both computational time and the number of iterations. In these conditions the method becomes competitive with existing computation codes (HFSS, MoMentum and others).

There are still several possibilities to improve convergence, such as the use of physically-based pre-conditioners. The first encouraging trials have taken place in the 1-D case [RAV 12]. Nevertheless for these studies, the modes TE and TM are isolated modes. They do not interact. The best physical pre-conditioners differ between the TE case and the TM case. In 3D, both types of mode coexist. The best physically-based pre-conditioners in 3D must therefore be a compromise solution.

The simulation for a particularly high number of frequencies, within the frequency band studied, proves to be interesting since the previous frequency solution may be used as an initial condition. However, the number of iterations is also directly linked to the number of frequencies. A more judicious choice of frequencies to simulate within the frequency band would therefore be interesting.

6

Application of WCIP to Diffraction Problems

6.1. Introduction

The study of electromagnetic wave diffraction problems arouses great interest among those working in the field of telecommunications. Indeed, we have seen the emergence and development of fresh demand for embedded systems for space applications, aviation systems, and cars (anti-collisions radars, tele-payment freeway systems, and other related devices), and for publicly used systems (satellite receivers, multi-point networks, mobile radio-communications, satellite tracking systems, and other similar devices), for example in the aviation sphere. This study is thus significant for modeling antennae through the use of radomes [ABD 01] as well as diffraction by jet engines (see [ROS 95, ANA 95, ANA 96]). The geometry of multilayer structures of any boundary shape is often the analytical solution for diffraction problems. This geometry induces competition from numerous numerical computational methods. Amongst the most popular methods or theories developed, we may cite the various integral or differential methods which are indeed considered to be high-performing methods, and which saw their heyday by modeling multiple structures. However, they still require too much time for simulation and sometimes remain difficult to formulate, in particular for high-complexity structures.

Chapter written by Noemen AMMAR, Taoufik AGUILI and Henri BAUDRAND.

In the first part of this chapter, we develop the Wave Concept Iterative Procedure (WCIP) method using cylindrical coordinates with the aim of resolving diffraction and electromagnetic wave penetration problems, by using multilayer cylindrical structures. The objective of this formulation is to determine a simple and efficient relationship which allows us to link both incident and reflected waves within the various media around the discontinuities by expressing the reflection within the modal sphere and diffraction, distinguished by boundary conditions and continuity conditions, within the spatial domain. The iterative process uses the Fourier transform and its converse in moving from one sphere to the other, through which it also provides both the aptitude and capacity for quick implementation.

In the second section, being based upon the development of the method in cylindrical coordinates, the WCIP method will be used for studying structures which can take any form. Some modifications are provided to the WCIP formulation so as to adapt the newly developed numerical method for the analysis of arbitrary geometrical objects.

6.1.1. *Diffraction by multilayer cylindrical structures*

6.1.1.1. *Theoretical formulation*

Figures 6.1 and 6.2 represent the general structure of an enclosed multilayer structure. Each dielectric layer is situated at a radius R_i and distinguished by its physical parameters (ε_i, μ_i). The intermediate layers which link waves, both incoming and outgoing, the two consecutive interfaces (Ω^i and Ω^{i+1}) are represented by quadrupoles $\hat{\Gamma}_{pass}^{(i-1,i)}$. The extreme layers are on the other hand treated as dipoles. The variables $\hat{\Gamma}_{\text{int}}$ and $\hat{\Gamma}_{ext}$ are reflections of endings expressing the shutdown conditions both above and below the structure. Continuity equations in terms of waves apply.

Upon each interface (Ω^i), both the metallic and the dielectric parts are defined by the angles $\Psi^{(j)}{}_{\mathrm{M}}{}^{(i)}$ and $\Psi^{(j)}{}_{\mathrm{D}}{}^{(i)}$, such that:

– $\Psi^{(j)}{}_{\mathrm{M}}{}^{(i)} = \Psi^{(i)}{}_{2j}{}^{(i)} - \Psi^{(i)}{}_{2j}{}^{(i)}{}_{-1}$, represent the angular length of the strip $j^{(i)}$;

– $\Psi^{(j)}{}_{\mathrm{D}}{}^{(i)} = \Psi^{(i)}{}_{2j}{}^{(i)}{}_{+1} - \Psi^{(i)}{}_{2j}{}^{(i)}$, represent the angular length of the slit between the strips $j^{(i)}$ and $j^{(i)}+1$. The index $j^{(i)}$ designates the strip labeled (j), which is placed across the interface (Ω^i).

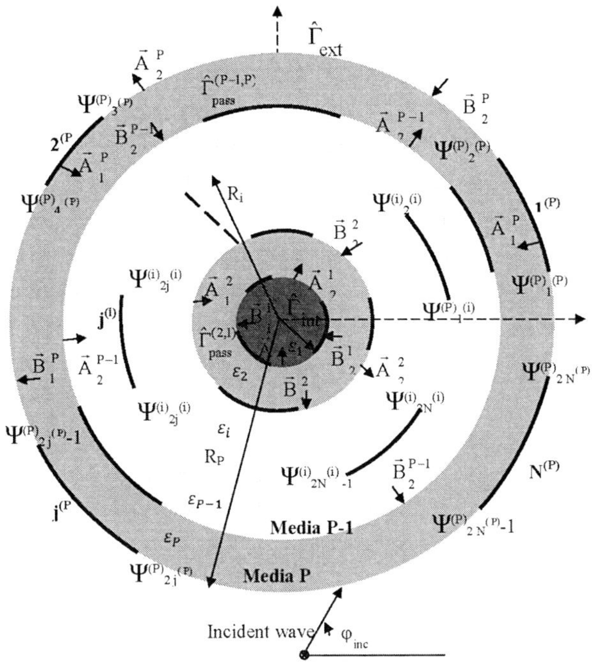

Figure 6.1. *Inhomogeneous multilayer structure*

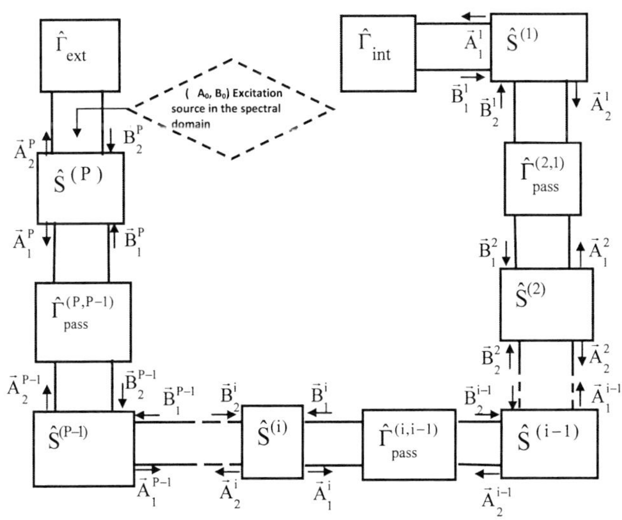

Figure 6.2. *Corresponding diagram shown in waves*

The pair (A_0, B_0) is all of the waves that are associated with the electromagnetic field. They are defined within the spectral domain.

Indices 2 and 1 indicate, respectively, the surfaces located directly above and below the discontinuity impedance Ω^i. $\vec{B}_q^i$ and $\vec{A}_q^i$ respectively represent incoming and outgoing waves, at interface level, with q being the values 1 or 2.

6.1.2. *Descriptors for spectral components of reflection operators*

In the spectral domain, two types of situations and therefore equations are possible by virtue of designing a multilayer system. Either it is the case that waves leaving discontinuity interfaces will be reflected by the environments surrounding these interfaces (either exterior or interior), or on the contrary, these waves give rise to those stimulating another neighboring interface. In the latter case, it is a question of determining the quadrupole connection between waves entering and leaving each of these interfaces.

To simplify the issue, let us consider the example of two cylindrical layers which are separated by interfaces. These three types of operators are defined within a two-cylinder coaxial system (Figure 6.3). They ensure that the relationships between waves accord with [6.1], [6.2] and [6.3].

$$\vec{B}_4 = \hat{\Gamma}_{ext} \vec{A}_4 \qquad [6.1]$$

$$\vec{B}_1 = \hat{\Gamma}_{\text{int}} \vec{A}_1 \qquad [6.2]$$

$$\begin{pmatrix} \vec{B}_2 \\ \vec{B}_3 \end{pmatrix} = \hat{\Gamma}_{pass} \begin{pmatrix} \vec{A}_2 \\ \vec{A}_3 \end{pmatrix} \qquad [6.3]$$

Taking account of the axial symmetry of the structure, the development of reflection operators within cylindrical modal bases is shown by the following general equation:

$$\hat{\Gamma}_\sigma = \sum_n \left| f_{n.R_i}(\phi) \right\rangle \Gamma_n^\sigma \left\langle f_{n.R_j}(\phi) \right| \qquad [6.4]$$

where $f_{n.R_i}(\phi)$ represents the modal base of a cylinder with a radius R_i, it is defined by:

$$f_{n.R_i}(\phi) = \frac{1}{\sqrt{2\pi R_i}} e^{j(n\phi)} \qquad [6.5]$$

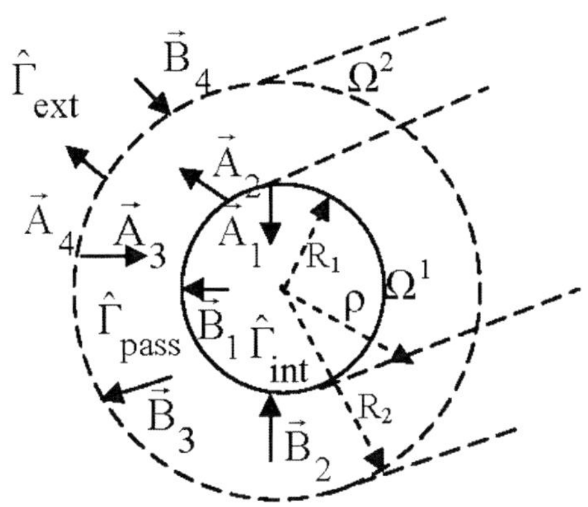

Figure 6.3. *Reflection operators for two coaxial cylinders*

Γ_n^σ represents the operator modal coefficients $\hat{\Gamma}_\sigma$. The index σ designates the symbols "int", "ext" or "pass".

NOTE.– In [6.4], the radii R_i and R_j are equal (for σ = int or ext but they are different for σ = pass).

6.1.3. *The modal coefficients Γ_n^{ext} and Γ_n^{int}*

The modal coefficients Γ_n^{ext} and Γ_n^{int} are deduced from the modal admittances by using the following relationship:

$$\Gamma_n^\sigma = (II - z_{0.i} Y_n^\sigma)(II + z_{0.i} Y_n^\sigma)^{-1} \qquad [6.6]$$

The index σ designates, in this case, either of the indices int or ext. $z_{0.i}$ represent the impedance of a surface i where they are defined waves.

For both polarizations TE and TM, the modal impedance matrices, which describe the reactions of spaces which are exterior and interior to a cylinder of radius R, are respectively expressed (see [RAV 03]) by:

– Hankel functions of the second kind and their by-products $H_n^2\left(k_\rho R\right)$ [6.7] and their by-products. These functions represent waves which radiate in the direction $+\rho$;

– Bessel functions of the first kind $J_n\left(k_\rho R\right)$ and their by-products [6.8]. They describe waves which are propagated within a circular waveguide:

$$Y_n^{ext} = \begin{pmatrix} \dfrac{k_\rho}{jw\mu}\dfrac{H_n^2\left(k_\rho R\right)}{H_n'^2\left(k_\rho R\right)} & 0 \\ 0 & \dfrac{jw\varepsilon}{k_\rho}\dfrac{H_n'^2\left(k_\rho R\right)}{H_n^2\left(k_\rho R\right)} \end{pmatrix} \quad [6.7]$$

$$Y_n^{\text{int}} = \begin{pmatrix} -\dfrac{k_\rho}{jw\mu}\dfrac{J\left(k_\rho R\right)}{J'\left(k_\rho R\right)} & 0 \\ 0 & -\dfrac{jw\varepsilon}{k_\rho}\dfrac{J'\left(k_\rho R\right)}{J\left(k_\rho R\right)} \end{pmatrix} \quad [6.8]$$

6.1.4. *Modal coefficients* Γ_n^{pass}

The modal coefficients Γ_n^{pass} are determined by knowledge of the modal components of electromagnetic fields, for both polarizations TE and TM, in the space generated by both cylinders (see [RAV 03]). In fact, the components z of the electric and magnetic fields are linear combinations of the Bessel functions of the first and second kind as below:

$$E_{z.n} = \left(a_{n.}^{TM} J_n(k_\rho \rho) + b_{n.}^{TM} Y_n(k_\rho \rho)\right)e^{jn\phi} \quad [6.9]$$

$$H_{z.n} = \left(a_{n.}^{TE} J_n(k_\rho \rho) + b_{n.}^{TE} Y_n(k_\rho \rho)\right)e^{jn\phi}$$

It is accepted that a_n^{TM} , b_n^{TM} , a_n^{TE} and b_n^{TE} are the individual values of E_z and H_z within cylindrical modal bases.

From Maxwell's equations, the modal components of the electric fields and current density are expressed across the surfaces of both cylinders according to the modal coefficients per the equations in [6.10] and [6.11].

$$\begin{pmatrix} E^{\phi}_{2.n} \\ E^{z}_{2.n} \\ E^{\phi}_{3.n} \\ E^{z}_{3.n} \end{pmatrix} = \begin{pmatrix} 0 & 0 & \frac{jw\mu}{k_\rho} J'_{n.1} & \frac{jw\mu}{k_\rho} Y'_{n.1} \\ J_{n.1} & Y_{n.1} & 0 & 0 \\ 0 & 0 & \frac{jw\mu}{k_\rho} J'_{n.2} & \frac{jw\mu}{k_\rho} J'_{n.2} \\ J_{n.2} & Y_{n.2} & 0 & 0 \end{pmatrix} \begin{pmatrix} a^{TE}_{n.} \\ b^{TE}_{n.} \\ a^{TM}_{n.} \\ b^{TM}_{n} \end{pmatrix} \qquad [6.10]$$

$$\begin{pmatrix} J^{\phi}_{2.n} \\ J^{z}_{2.n} \\ J^{\phi}_{3.n} \\ J^{z}_{3.n} \end{pmatrix} = \begin{pmatrix} 0 & 0 & J_{n.1} & Y_{n.1} \\ \frac{jw\varepsilon}{k_\rho} J'_{n.1} & \frac{jw\varepsilon}{k_\rho} Y'_{n.1} & 0 & 0 \\ 0 & 0 & -J_{n.2} & -Y_{n.2} \\ -\frac{jw\varepsilon}{k_\rho} J'_{n.2} & -\frac{jw\varepsilon}{k_\rho} Y'_{n.2} & 0 & 0 \end{pmatrix} \begin{pmatrix} a^{TE}_{n.} \\ b^{TE}_{n.} \\ a^{TM}_{n.} \\ b^{TM}_{n} \end{pmatrix} \qquad [6.11]$$

$J_{n.p}$ and $Y_{n.p}$ are Bessel functions of the first and second kind ($p = 1$ or 2).

From wave definition relationships, modal components of waves defined across the surfaces of both cylinders (Figure 6.3) are determined according to their modal coefficients using the following equations:

$$\begin{pmatrix} A^{\phi}_{2.n} \\ A^{z}_{2.n} \\ A^{\phi}_{3.n} \\ A^{z}_{3.n} \end{pmatrix} = \underbrace{\begin{pmatrix} jKJ'_{n.1} + J_{n.1} & jKY'_{n.1} + Y_{n.1} & 0 & 0 \\ 0 & 0 & J_{n.1} + jKJ'_{n.1} & Y_{n.1} + jKY'_{n.1} \\ jKJ'_{n.2} - J_{n.2} & jKY'_{n.2} - Y_{n.2} & 0 & 0 \\ 0 & 0 & J_{n.2} - jKJ'_{n.2} & Y_{n.2} - jKY'_{n.2} \end{pmatrix}}_{N} \begin{pmatrix} a^{TE}_{n.} \\ b^{TE}_{n.} \\ a^{TM}_{n.} \\ b^{TM}_{n} \end{pmatrix} \qquad [6.12]$$

$$\begin{pmatrix} B^{\phi}_{2.n} \\ B^{z}_{2.n} \\ B^{\phi}_{3.n} \\ B^{z}_{3.n} \end{pmatrix} = \underbrace{\begin{pmatrix} jKJ'_{n.1} - J_{n.1} & jKY'_{n.1} - Y_{n.1} & 0 & 0 \\ 0 & 0 & J_{n.1} - jKJ'_{n.1} & Y_{n.1} - jKY'_{n.1} \\ jKJ'_{n.2} + J_{n.2} & jKY'_{n.2} + Y_{n.2} & 0 & 0 \\ 0 & 0 & J_{n.2} + jKJ'_{n.2} & Y_{n.2} + jKY'_{n.2} \end{pmatrix}}_{M} \begin{pmatrix} a^{TE}_{n.} \\ b^{TE}_{n.} \\ a^{TM}_{n.} \\ b^{TM}_{n} \end{pmatrix} \qquad [6.13]$$

Taking the formula:

$$K = \frac{w\varepsilon z_0}{k_\rho} = \frac{w\mu}{k_\rho z_0}$$

ε and μ are respectively the permittivity and the permeability of the space between two cylinders and k_ρ is the radial propagation constant.

Modal components of the transition operators are deduced from both a reversal and a multiplication of both matrices M and N.

$$\Gamma_n^{pass} = M * N^{-1} = \begin{pmatrix} T_n^{11} & 0 & T_n^{12} & 0 \\ 0 & -T_n^{11} & 0 & T_n^{12} \\ T_n^{21} & 0 & T_n^{22} & 0 \\ 0 & T_n^{21} & 0 & -T_n^{22} \end{pmatrix} \qquad [6.14]$$

After standardization and establishing the power conservation, standardized modal coefficients are given by the equation in [6.15].

$$\Gamma_{n.N}^{pass} = \begin{pmatrix} T_n^{11} & 0 & \sqrt{\frac{R_1}{R_2}}T_n^{12} & 0 \\ 0 & -T_n^{11} & 0 & \sqrt{\frac{R_1}{R_2}}T_n^{12} \\ \sqrt{\frac{R_2}{R_1}}T_{n1}^{21} & 0 & T_n^{22} & 0 \\ 0 & \sqrt{\frac{R_2}{R_1}}T_n^{21} & 0 & -T_n^{22} \end{pmatrix} = \begin{pmatrix} \Gamma_n^{11} & 0 & \Gamma_n^{12} & 0 \\ 0 & -\Gamma_n^{11} & 0 & \Gamma_n^{12} \\ \Gamma_{n1}^{21} & 0 & T_n^{22} & 0 \\ 0 & \Gamma_n^{21} & 0 & -\Gamma_n^{22} \end{pmatrix} \qquad [6.15]$$

These modal coefficients will be applied later in our study to ensure wave transitions between two consecutive cylindrical layers.

6.1.5. *Spatial diffraction operator*

This operator generalizes the boundary conditions and the continuity conditions across the various domains making up a discontinuity interface of an impedance Ω^i. Its expression flows directly from descriptors of boundary conditions and electromagnetic field continuity at the level of various interfaces of the cavity. It is expressed by the following relationship:

$$\begin{bmatrix} \vec{A}_2^i \\ \vec{A}_1^i \end{bmatrix} = \hat{S}^i \cdot \begin{bmatrix} \vec{B}_2^i \\ \vec{B}_1^i \end{bmatrix} = \begin{pmatrix} -\frac{\eta_i^2 - 1}{\eta_i^2 + 1} H_D^i - H_M^i & \frac{2\eta_i^2}{\eta_i^2 + 1} H_D^i \\ \frac{2\eta_i^2}{\eta_i^2 + 1} H_D^i & \frac{\eta_i^2 - 1}{\eta_i^2 + 1} H_D^i - H_M^i \end{pmatrix} \begin{bmatrix} \vec{B}_2^i \\ \vec{B}_1^i \end{bmatrix} \quad [6.16]$$

With the formula:

$\eta_i = \frac{\sqrt{z_{0i}}}{\sqrt{z_{0i-1}}}$, such that

z_{0i} is the intrinsic impedance of the material (i). It is stated according to the impedance of the vacuum z_0, by

$$z_{0i} = \frac{z_0}{\sqrt{\varepsilon_{ri}}}$$

$-\varepsilon_{ri}$: the permittivity relative to the dielectric property of the material i.

We may recall that:

$-H_D^i$ is a Heaviside unit step defined such that:

$$H_D^i = \begin{cases} 1 & \textit{On the dielectric surface of the given layer } (i) \\ 0 & \textit{elsewhere} \end{cases}$$

$-H_M^i$ is a Heaviside unit step defined such that:

$$H_M^i = \begin{cases} 1 & \textit{On the metal surface of the layer } (i) \\ 0 & \textit{elsewhere} \end{cases}$$

6.1.6. *Excitation source*

Concerning the diffraction problem, the excitation source considers a wave plane, that is to say the incident electromagnetic field. The electric and magnetic fields which make it up will be defined at the interface level of the upper layer. They are taken into account within equations which govern the relationship between waves in the spectral domain. From the initial tangential fields, denoted by the use of $\vec{E}^{inc}$ and $\vec{J}^{inc}$ $(\vec{J}^{inc} = \vec{H}^{inc} \wedge \vec{n})$, we can compute an incident wave (B_0) by the wave definition. For both polarizations (TM and TE), the incident electric fields are expressed by cylindrical coordinates according to the equations in [6.17] and [6.18].

$$\vec{E}_z^{inc} = E_0 \sum_{n=-\infty}^{+\infty} i^{-n} J_n(k_0 \rho) \exp(i(n(\phi - \phi_{inc}))\vec{e}_z \qquad [6.17]$$

$$\vec{E}_\phi^{inc} = E_0 \sum_{n=-\infty}^{n=+\infty} i^{-n+1} J_n'(k_0 \rho) e^{jn(\phi-\phi_{inc})} \vec{e}_\phi \qquad [6.18]$$

$$k_0 = w\sqrt{\mu_0 \varepsilon_0}$$

φ_{inc} is the phase difference angle of the incident field. It is measured using a phase difference of π in relation to it.

6.1.7. *Iterative process*

According to the corresponding diagram (Figure 6.4), it is possible to end up at the iterative process which is governed by two systems of equations; one system being written in the spatial domain and the other in the modal domain. The transition from one domain to another happens by the intermediary of the Fourier transform and its converse. This recursive relationship is capable of determining the incident and reflected wave at any given iteration (k) from initial waves. Electromagnetic fields, current density and electric field are determined at each iteration. The iterative process is halted when the electric field (or current density) converges.

6.2. Application

Three types of structures are proposed during the simulation. The first two structures concern the issue of diffraction by two single-layer cylindrical structures, the two others are devoted to the study of diffraction, and indeed by two cylindrical layers.

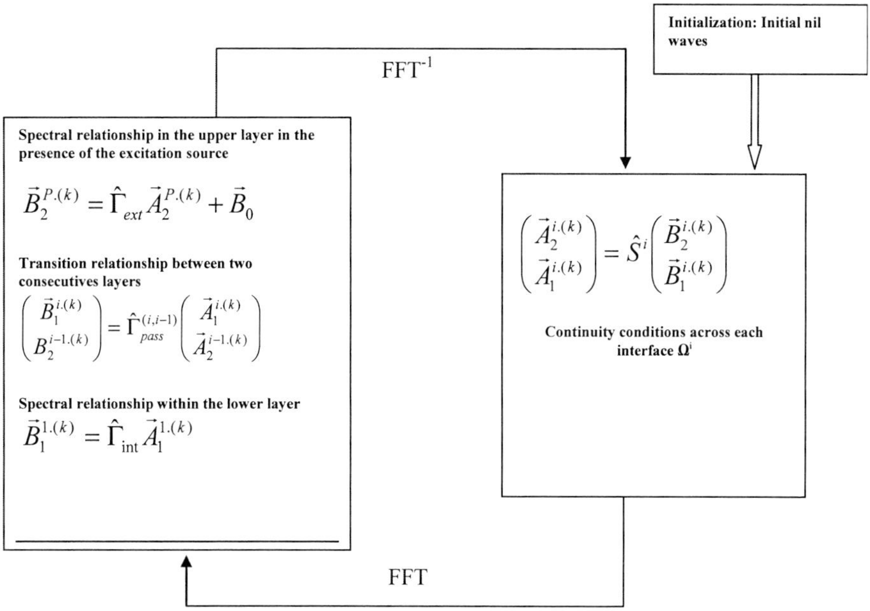

Figure 6.4. *Diagram of iterative process*

6.2.1. *Dielectric cylinder diffraction*

We will now consider the case of a dielectric cylinder with a permittivity ε_r, excited by a wave plane. We are particularly interested in the distribution of electromagnetic fields at the level of the discontinuity surface and also the diffracted fields in the remote zone locations.

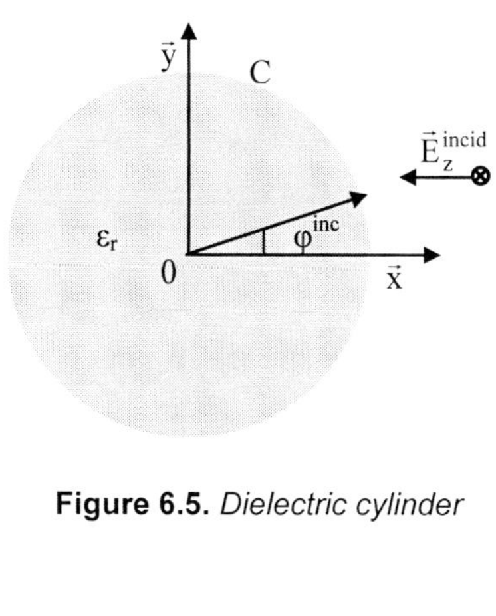

Figure 6.5. *Dielectric cylinder*

Within the simulation, the circular contour is described with a mesh of 180 pixels, the elementary dimension of which following the direction $\vec{\varphi}$ is given by $\Delta\varphi = 2°$. The simulation includes all of the geometrical characteristics that are indicated in Figure 6.6(a). Figure 6.6(b) shows the conditions under which the simulation was carried out.

Geometric characteristics:

R = 0.3*λ and R = λ

Δϕ = 2°

a)

Simulation constraints:

TM incident wave with $\phi_{inc} = 0$

Number of iterations (kit):50

b)

Figure 6.6. *Table of simulation constraints*

The results in Figure 6.7 show that the convergence of both the actual and hypothetical parts of the field E_z for both cylinder rays is obtained fairly quickly, at around 20 iterations.

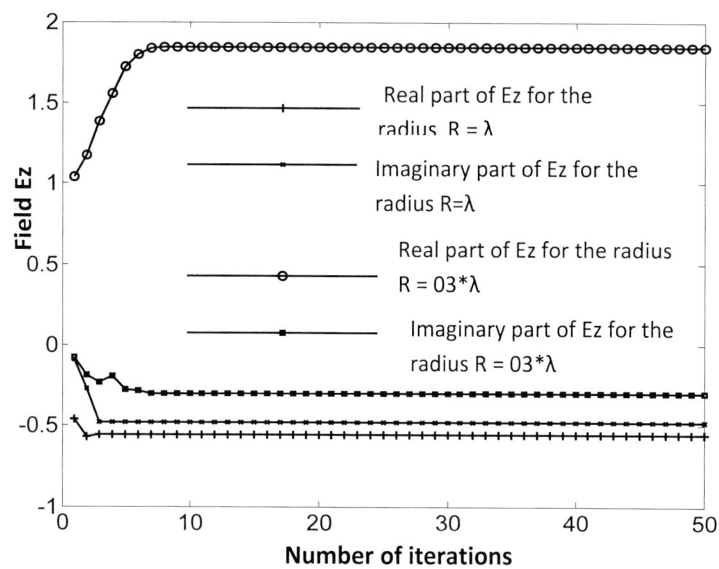

Figure 6.7. *Convergence of an electric field*

At the discontinuity surface, the distribution of the electric field for both radii is the formula (R = 0.3λ and R = λ) which is given by Figures 6.8 and 6.9. It is clear that for radii which are comparable to the wavelength, the diffracted field will display oscillations. This is the resonance phenomenon.

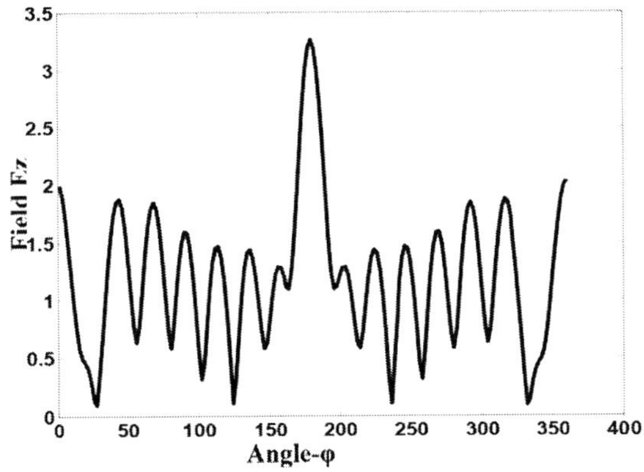

Figure 6.8. *Electric field modulus for R = λ*

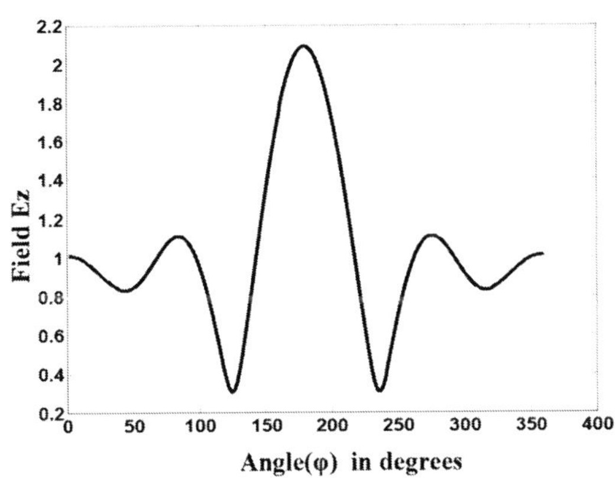

Figure 6.9. *Electric field modulus for R = 0.3λ*

To determine the field diffracted to infinity, after the convergence of the electric field across the cylinder surface, the transition operator defined in the first section is applied to the wave $\vec{A}$ emitted from the surface of the cylinder.

In using this application, we will calculate the radar echo value, which is the relationship of square modules of diffracted and incident fields measured across a semi-circle which is located in a highly remote location in relation to the structure concerned.

$$\mathrm{RCS} = \lim_{\rho \to \infty} 2\pi\rho \frac{\left|E^{\mathrm{diff}}\right|^2}{\left|E^{\mathrm{inc}}\right|^2} \qquad [6.19]$$

Figures 6.10 and 6.11 show the variation of the radar cross-section (RCS) coefficient for the rays $R = \lambda$ and $R = 0.3\lambda$ and for an incidence $\varphi_{\mathrm{inc}} = \pi$.

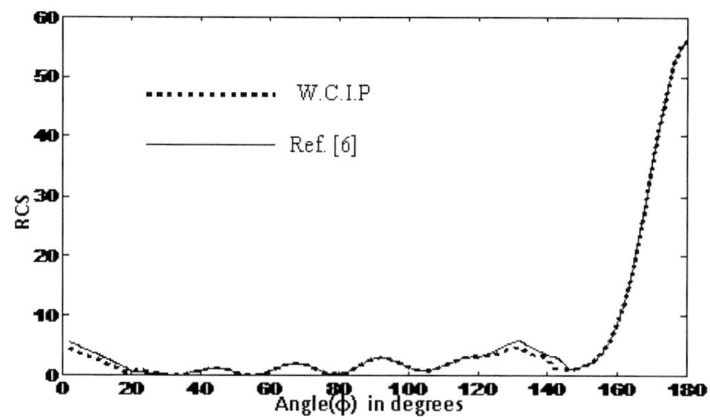

Figure 6.10. *RCS variation for R = λ*

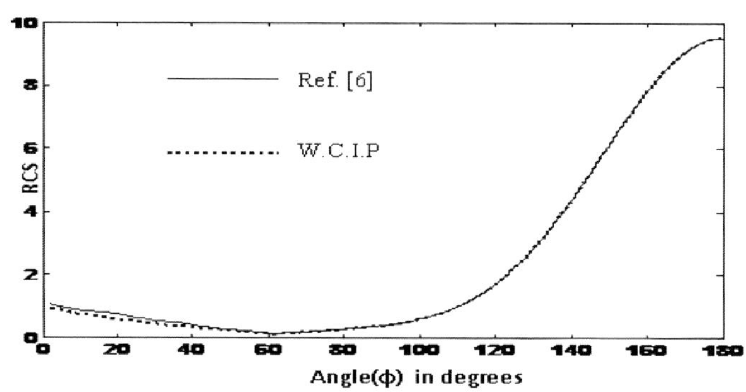

Figure 6.11. *RCS variation for R = 0.3λ*

6.2.2. *Diffraction by metallic strips*

The examples covered in this section consider metallic strips in cylindrical form, by considering two strip configurations:

– first case study: $\{\psi_1, \psi_2, \psi_3, \psi_4\} = \{25°, 155°, 205°, 335°\}$;

– second case study $\{\psi_1, \psi_2, \psi_3, \psi_4\} = \{60°, 120°, 240°, 360°\}$.

The incident field produces a phase difference of an angle $\varphi_{inc} = -90°$ compared with the horizontal straight line (Figure 6.12).

Figures 6.13, 6.14, 6.15 and 6.16 show the electric field and the current density for both geometric configurations, obtained after 120 iterations. Let us now examine the curves given by these previous figures on which it may be noticed that the more the slit dimensions increase, the weaker the electromagnetic wave penetration. The electric field modulus demonstrates significant alignment with that shown in works in the field [YIN 02, YEO 02]. The boundary conditions are well established.

Within this particular application, we consider the case of three geometric metallic strips which is given by:

$\{\psi_1, \psi_2, \psi_3, \psi_4, \psi_5, \psi_6\} = \{30°, 90°, 150°, 210°, 270°, 330°\}$

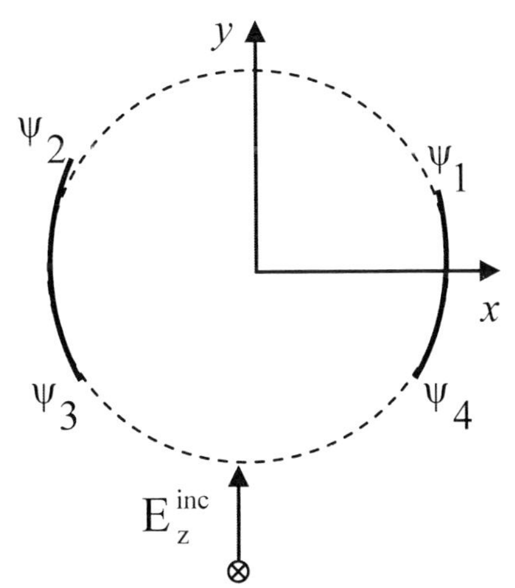

Figure 6.12. *Diffraction by two metallic strips*

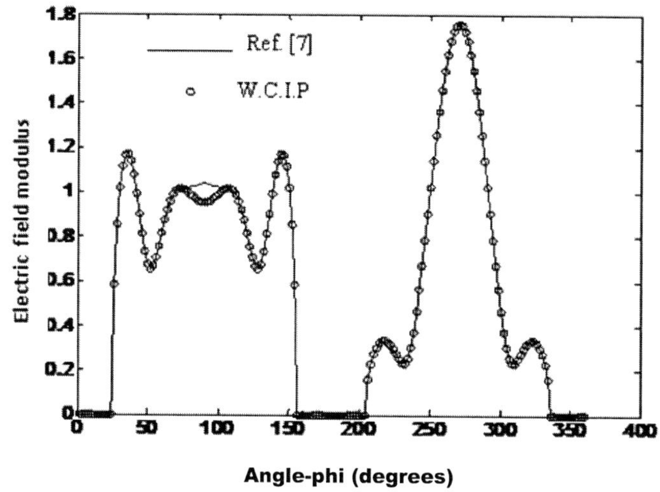

Figure 6.13. *Electric field modulus in p = R for {ψ_1, ψ_2, ψ_3, ψ_4} {25°, 155°, 205°, 335°}*

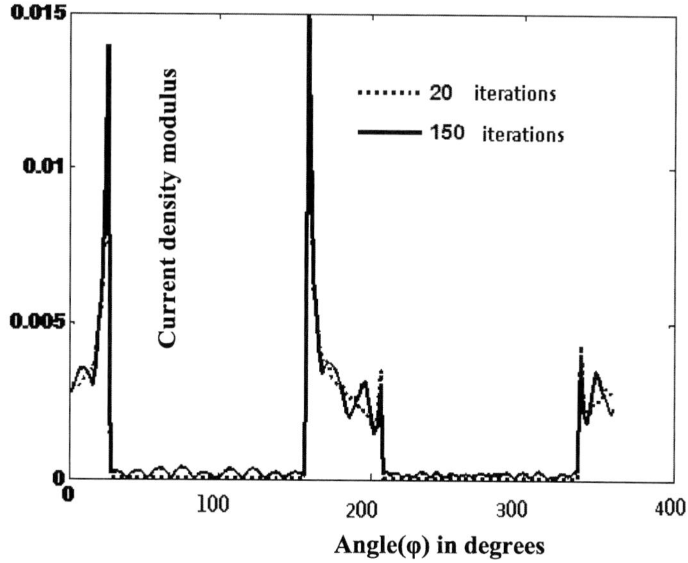

Figure 6.14. *Current density modulus in p = R for {ψ_1, ψ_2, ψ_3, ψ_4} = {25°, 155°, 205°, 335°}*

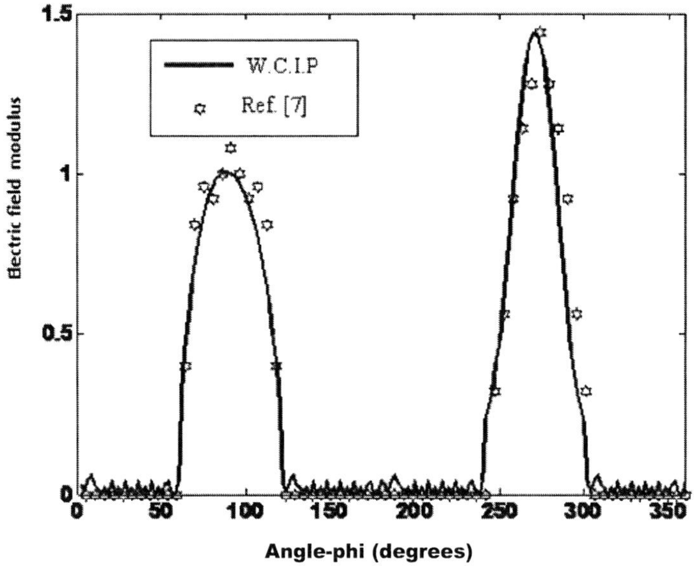

Figure 6.15. *Electric field modulus in $p = R$ for $\{\psi_1, \psi_2, \psi_3, \psi_4\} = \{60°, 120°, 240°, 360°\}$*

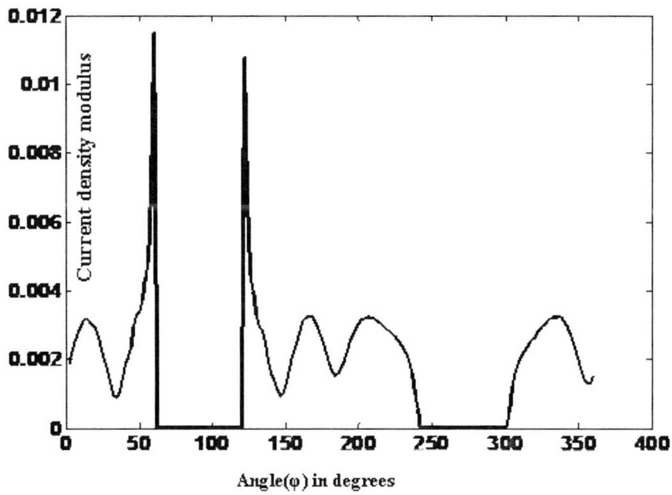

Figure 6.16. *Current density modulus in $p = R$ for $\{\psi_1, \psi_2, \psi_3, \psi_4\} = \{60°, 120°, 240°, 360°\}$*

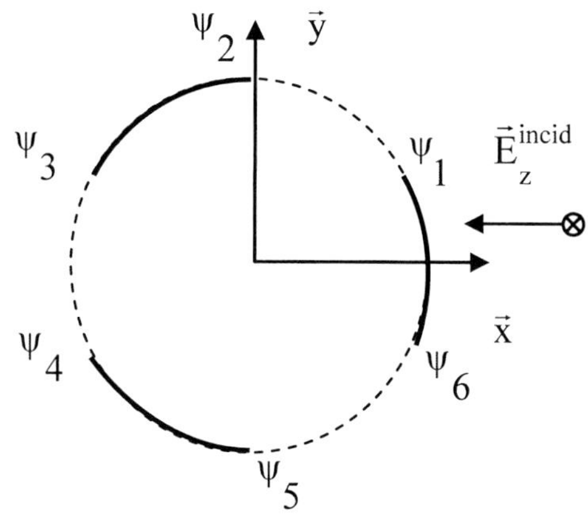

Figure 6.17. *Diffraction by three cylindrical metallic strips*

The structure is excited by a wave plane TM, with an angle of incidence $\varphi_{\text{inc}} = \pi$.

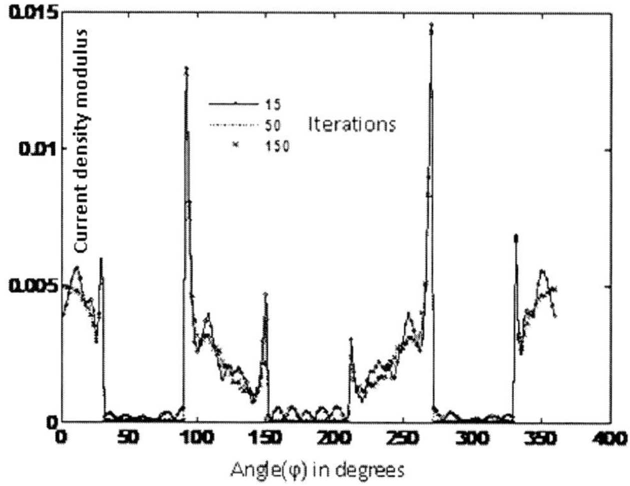

Figure 6.18. *The current density modulus*

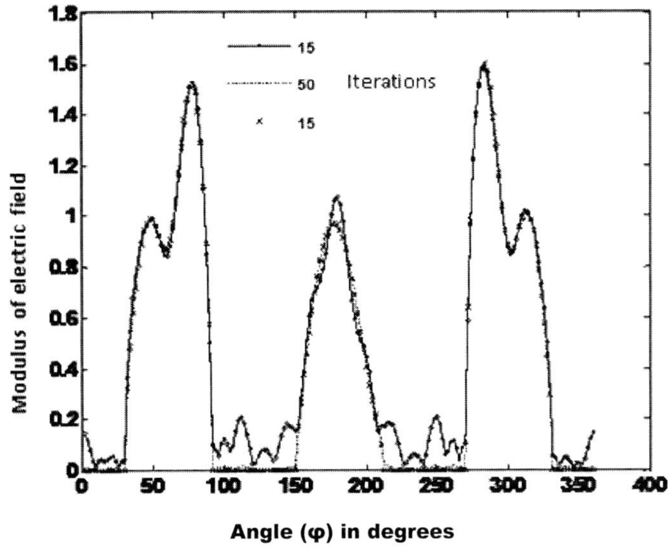

Figure 6.19. *The electric field modulus*

Observation of the electric field and the current density is able to confirm electromagnetic hypotheses, that is to say that current density reaches its maximum on the edges of the magnetic strips when it is nil on the dielectric parts of the apparatus. In addition, the electric field establishes the boundary conditions.

Subsequently, the distribution of the electromagnetic field within the space surrounding the cylindrical strips is shown. The mesh density is fixed at a 20-pixel wavelength in the direction of $\vec{\rho}$ and $\frac{2\pi}{180}$ in the direction of $\vec{\varphi}$. The beam of cylinder (C) is fixed at $R_C = \lambda$, the electric fields are shown in 3D for a beam ρ, which varies from zero to 2λ.

Two simulations have been carried out by modifying the positions, the number and the lengths of metallic strips:

– first case study: $\{\psi_1, \psi_2, \psi_3, \psi_4\} = \{25°, 155°, 205°, 335°\}$;

– second case study: $\{\psi_1, \psi_2, \psi_3, \psi_4, \psi_5, \psi_6\} = \{30°, 90°, 150°, 210°, 270°, 330°\}$.

For both of these configurations we will consider a wave plane TM with a phase difference $\varphi_{\text{inc}} = \pi$.

The numerical results in Figures 6.20 and 6.21 show that the boundary conditions are established by the modulus of the electric field and that the laws of electromagnetism are adhered to. Indeed, outside of the confines of cylinder C, the module of diffracted electric fields varies according to $1/\rho$.

6.2.3. *Coaxial multi-strip structure*

We should consider as an excitation source the plane waves TE, the expression of which, the component E_φ in the electric field, is shown by using the formulae below.

In this application we use: $\varphi_{\text{inc}} = -\dfrac{\pi}{4}$.

Figures 6.22, 6.23 and 6.24 show the convergence results of both the electric field and the current density according to the configuration iterations. The convergence of these electromagnetic magnitudes is obtained in fewer than 40 iterations.

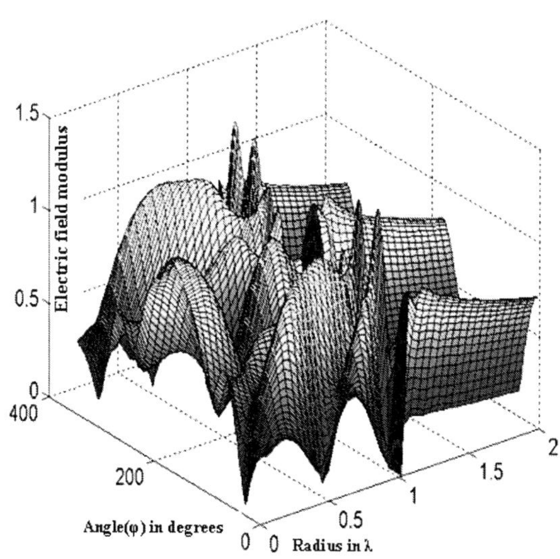

Figure 6.20. *Modulus of the electric field for* $\{\psi_1,\psi_2,\psi_3,\psi_4,\psi_5,\psi_6\} = \{25°,155°,205°,335°\}$

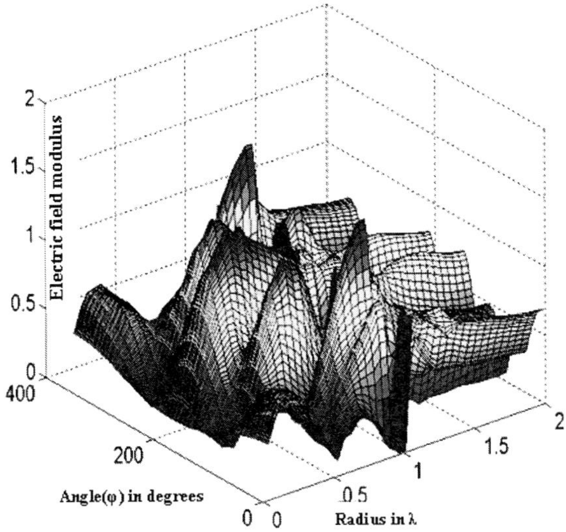

Figure 6.21. *Modulus of the electric field for* $\{\psi_1, \psi_2, \psi_3, \psi_4, \psi_5, \psi_6\} = \{30°, 90°, 150°, 210°, 270°, 330°\}$

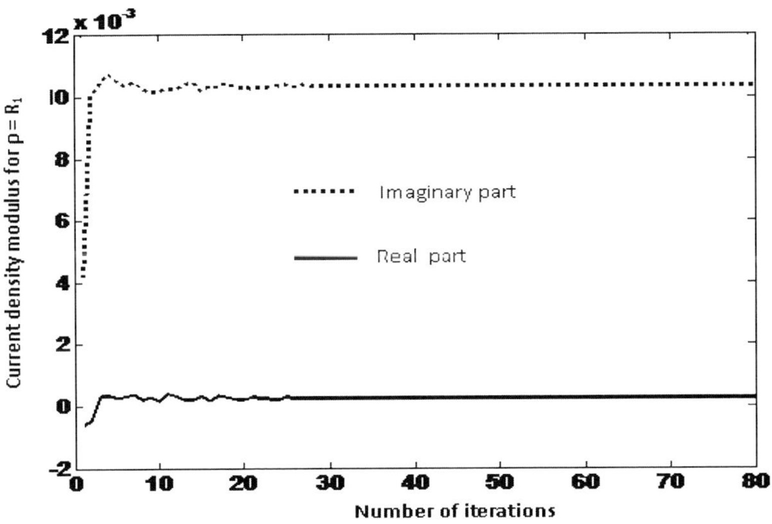

Figure 6.22. *Convergence of current density across the cylinder of the radius* R_1

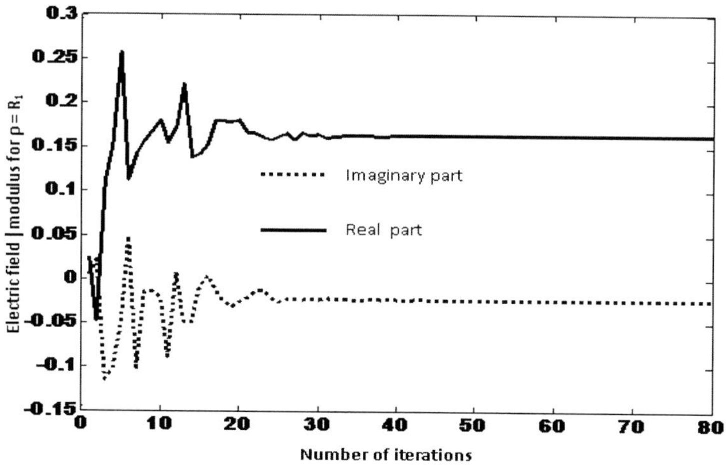

Figure 6.23. *Convergence of the electric field across the cylinder of the beam R_1*

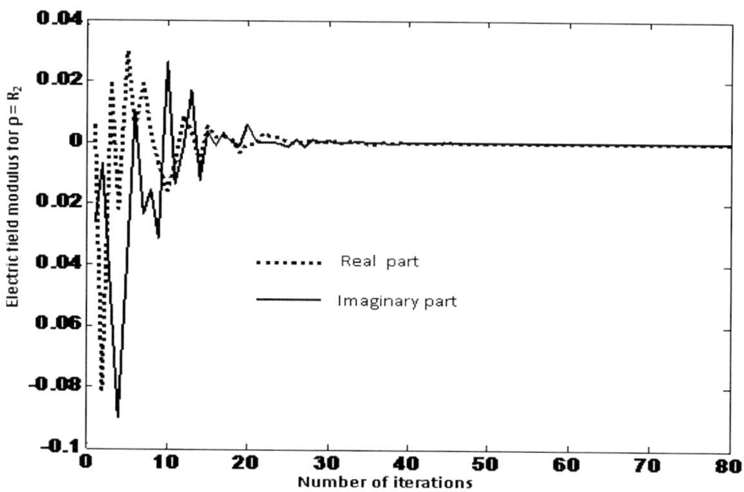

Figure 6.24. *Convergence of the electric field the cylinder of the beam R_2*

Figures 6.25 and 6.26 show the distribution of the electric field and the current density for the geometric configuration A. Upon examining the curves of these figures, we may note that for an incidence angle $\varphi_{\text{inc}} = -90°$, contrary to the electric field, the current density is more significant across the strips of the interior cylinder than across that of the exterior cylinder.

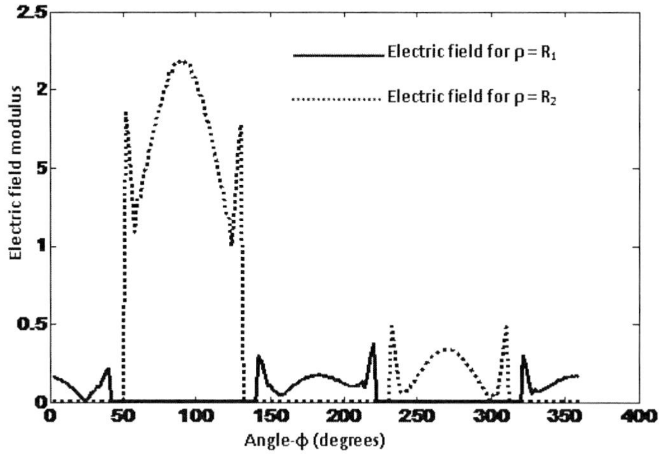

Figure 6.25. *Electric field for configuration A*

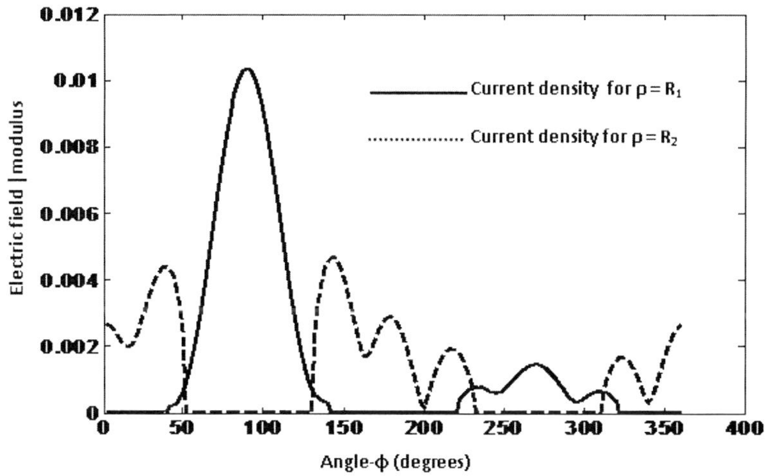

Figure 6.26. *Current density for configuration A*

The same comments may be made for the configuration B (Figures 6.27 and 6.28). Indeed, the electric field is more important at the level of the cylinder apertures of beam R_2 than those of the cylinder of beam R_1. The boundary conditions are well established for both configurations.

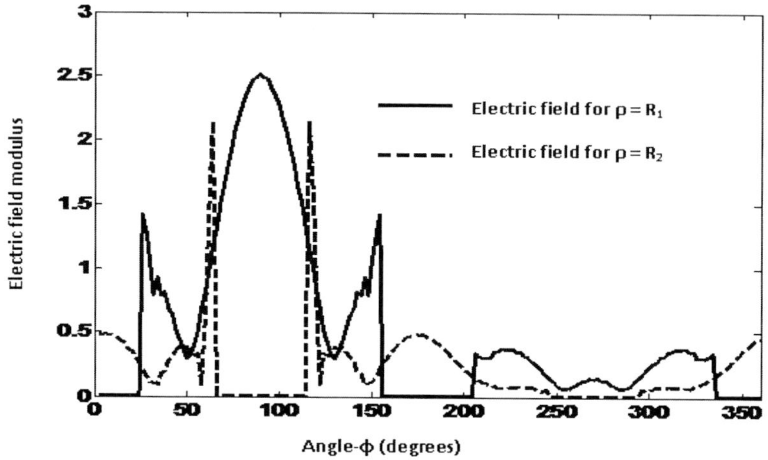

Figure 6.27. *Electric field for configuration B*

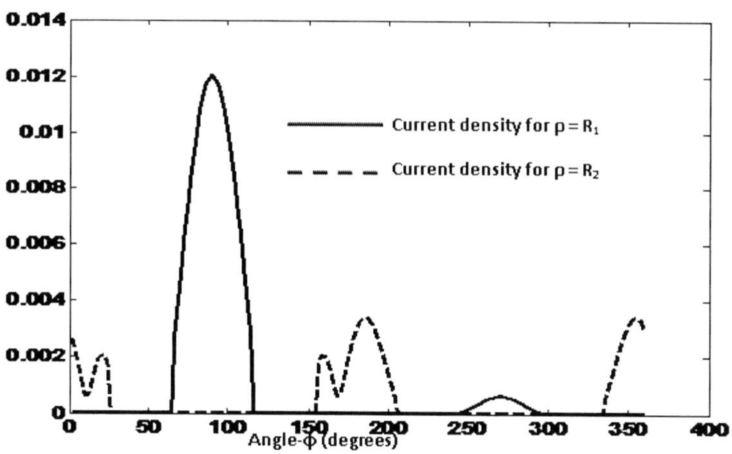

Figure 6.28. *Current density for configuration B*

Next, based upon what has been established in the previous section, the electric fields and the current density within the structure are determined for both geometric configurations (A and B). The surface containing the microwave structure is digitized with a mesh density of 180 pixels per the $\vec{\varphi}$ direction and 25 pixels per the $\vec{\rho}$ direction.

The representation of two dimensions of both the electric field and current densities for both configurations (Figures 6.29, 6.30, 6.31 and 6.32), indicates that the penetration of the electromagnetic field within the cavity becomes weak if the size of the metallic strips increases.

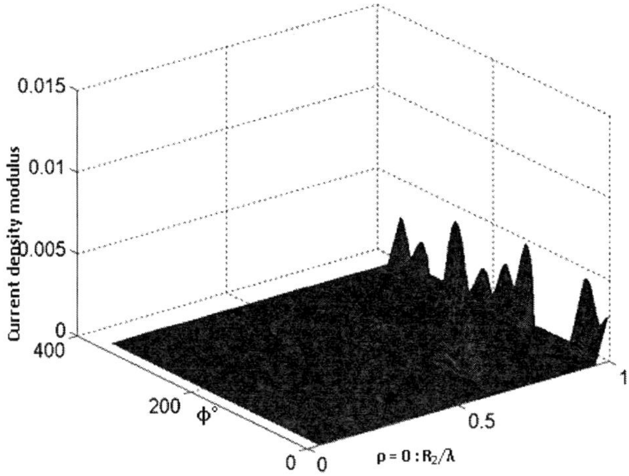

Figure 6.29. *Current density for configuration A*

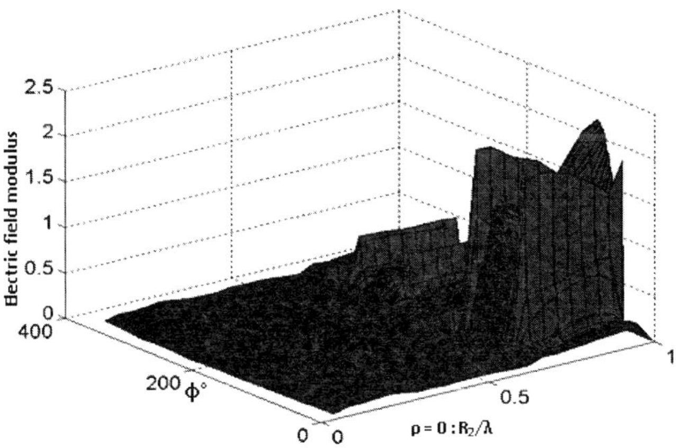

Figure 6.30. *Electric field for configuration A*

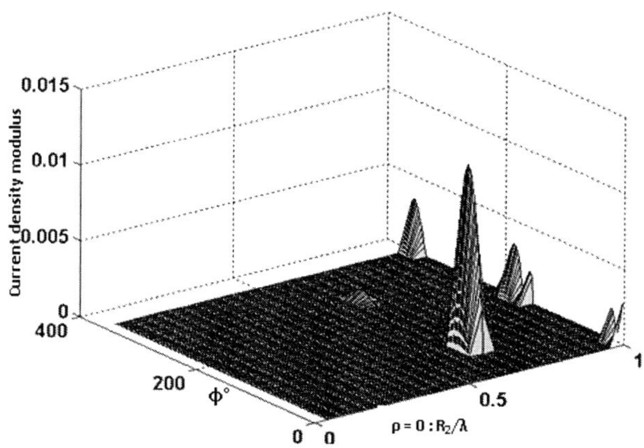

Figure 6.31. *Current density for configuration B*

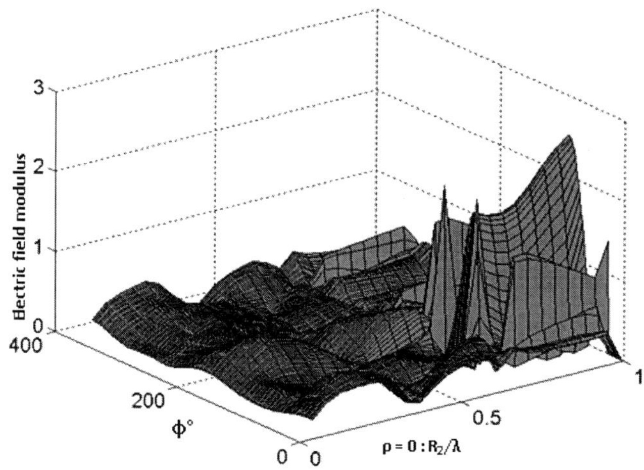

Figure 6.32. *Electric field for configuration B*

Upon resuming the geometric configuration A, by fixing the beam of the interior cylinder R_1 at $0.9*R_2$, Figures 6.33 and 6.34 show the distribution of the electric field and the current density for an incidence angle $\varphi_{\text{inc}} = -90°$.

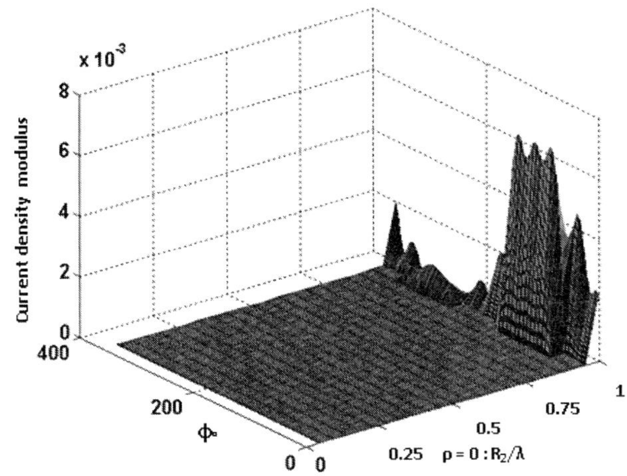

Figure 6.33. *Current density for configuration A using the formula R_1 = 0.9λ*

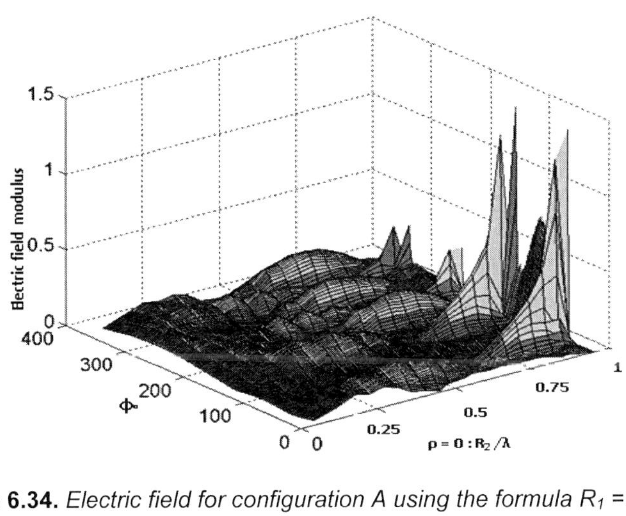

Figure 6.34. *Electric field for configuration A using the formula R_1 = 0.9*R_2*

For both R_1 values (R_1 = (2/3)*R_2 and R_1 = 0.9*R_2), the comparison of results given for the geometric configuration A shows that if the difference in the radii of the two cylinders (both internal and external) tends towards zero, the penetration of electromagnetic energy to the interior of the cavity becomes increasingly weaker.

6.2.4. *Diffraction by two dielectric co-axials*

Here, we are concerned with the diffraction of a plane wave TE by a cylindrical structure placed within free space, made up of two dielectric layers (see [AMM 08, AGU 08, BAU 08]), each layer i (i = 1, 2) is characterized by its relative dielectric constant (ε_{ri}) (Figure 6.35).

For this structure, we seek to compute both fields within close proximity and fields which have been diffracted in remote location zones by using the wave formulation. Geometric conditions are shown in Figure 6.36.

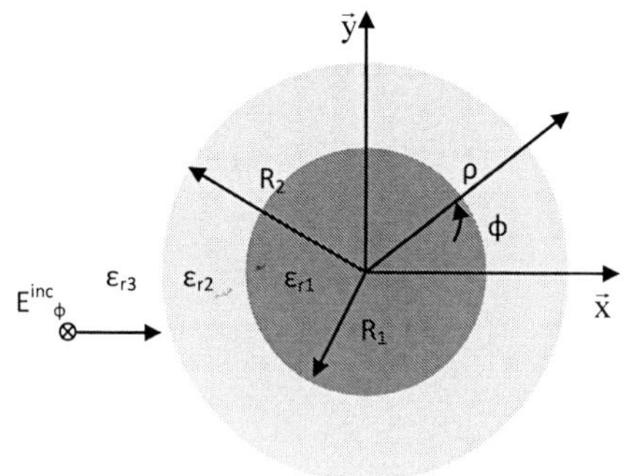

Figure 6.35. *Two-layer structure*

The incident wave is expressed as per that shown in [6.18]. Within this application we use the equation φ_{inc} = 0°.

Geometric Characteristics

$R_2 = 2.5\lambda$; $R_1 = 2\lambda$;

Mesh density $\Delta\phi = 2°$

Figure 6.36. *Geometric characteristics of Figure 6.36*

Figures 6.37 and 6.38 show the results of the convergence of the electric field across both discontinuity interfaces according to iterations. The curves show that electric field convergence is reached in less than 50 iterations.

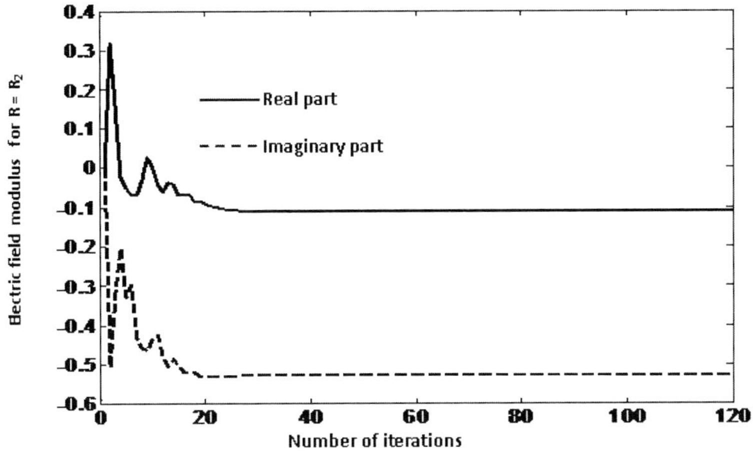

Figure 6.37. *Convergence of the electric field across the exterior interface*

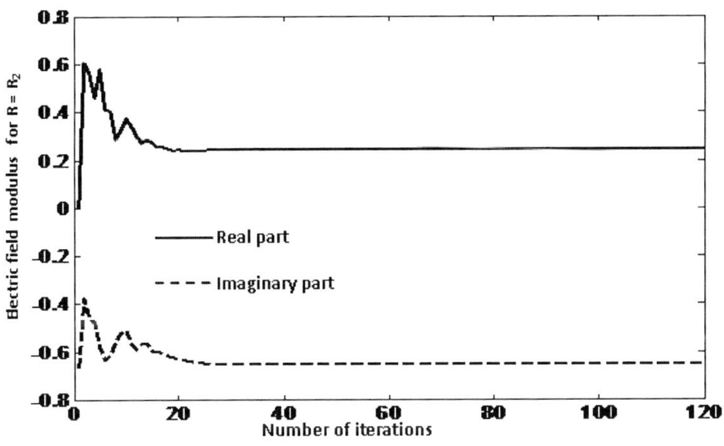

Figure 6.38. *Convergence of the electric field across the interior interface*

With the convergence of the electric field across both discontinuity interfaces, the infinite transition operator is applied to the surface of the cylinder for the radius R_2 to transform infinite waves $\vec{A}_2^2$ (see Figure 6.39) with the aim of computing the component E_φ of the electric field which is diffracted to infinity.

Figure 6.40 shows the electric field variation standardized in relation to the incident field, for $\varphi \in [-45°,45°]$ with an angle of incidence $\varphi_{in} = 0°$, the simulation results show that the electric field diffracted in a remote zone is at the maximum point for $\varphi = 0°$ and there is a close alignment with the reference formula (see [TSA 07]).

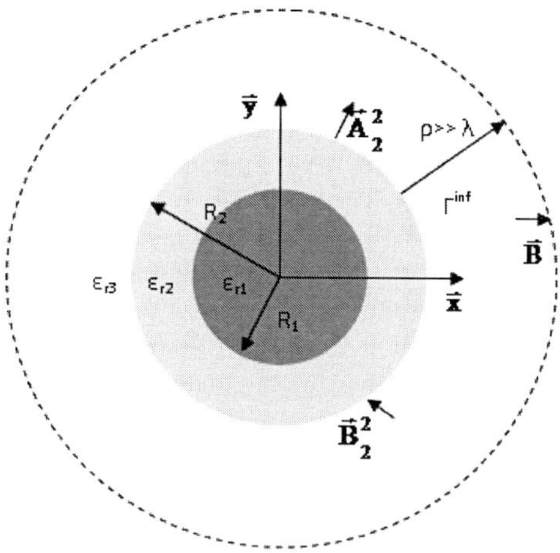

Figure 6.39. *Computation of infinite field*

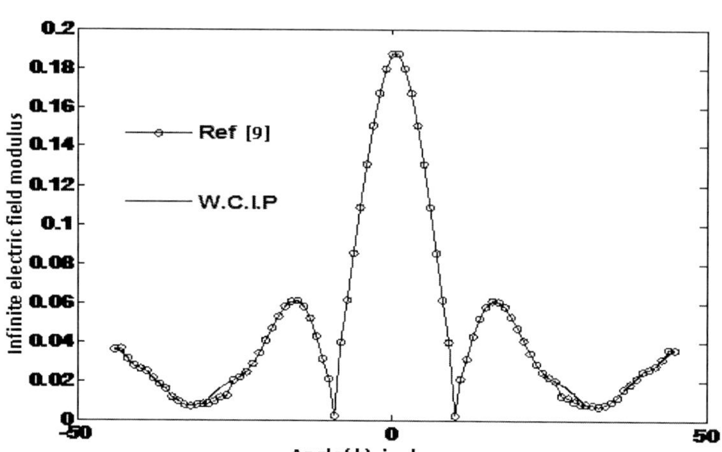

Figure 6.40. *Infinite electric field*

6.2.5. *Diffraction by structures of any shape*

To resolve problems involving structures of any form with the given wave formulation (Figure 6.41), the reaction of spaces around the surface under consideration should be distinguished by operators which are defined within the spatial domain so as to retain the shape of the structure.

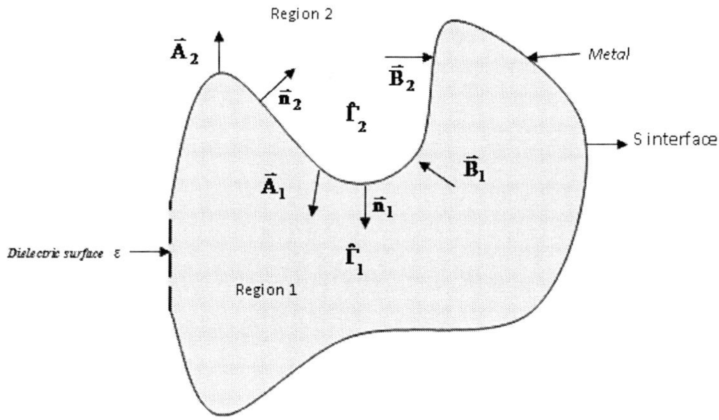

Figure 6.41. *Cross-section of a structure of any given form*

Figure 6.41 shows a transverse section of a structure with any non-homogeneous shape. Waves $\vec{A}_i$ and $\vec{B}_i$ are defined within both regions of the structure. The coupling between these waves is characterized by the operators $\hat{\Gamma}_i$, with *i* having either values 1 or 2.

To determine the operators $\hat{\Gamma}_i$, the interface of the structure should be discretized with a sufficiently fine mesh to obtain pixels (segments) which are regular in shape (Figure 6.42). The representative operator of the exterior or interior media on the (*S*) interface depends on an understanding of the interaction $I_{i,j}$ between each coupling of pixels within the structure. It is also necessary to take into account the interactions with the pixels representing the dielectric part of the structure. In fact, the incident wave across pixel P_i, results from the total of all waves emitted by all of the other pixels of the structure. The waves which are both emitted and received which cause the interaction, are placed and orientated in any way whatsoever within the space. Within this characterization no given source is considered. Diffracted waves which are received across a pixel are expressed according to waves emitted from other pixels of the structure.

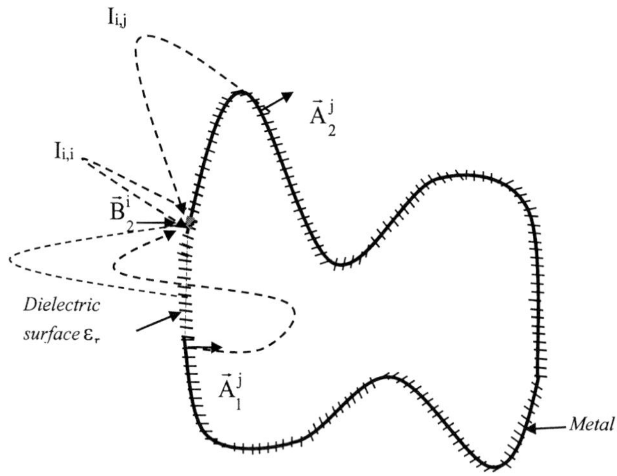

Figure 6.42. *Interaction between two pixels of the structure*

6.3. Coupling simulation

To distinguish the coupling features, it is first necessary to determine the modal base in which the diffraction operators are developed. Next, we determine the different specific modal diffraction operators which apply to each type of coupling.

The choice of modal base is necessary for the development of diffraction operators within the modal domain. This modal base depends upon the coordinates system in which propagation equations are expressed, i.e. cylindrical or spherical Cartesian coordinates. This system of coordinates should be capable of integrating two pixels whatever their positions and their orientations in the space (Figure 4.2). A given pixel is characterized by both a segment and a norm. To determine a common modal base with these two pixels, the latter should imperatively be included within one or two fictitious surfaces. These may be a single plane surface or two in parallel, and one or two co-axial cylinders or one or two concentric spheres.

The system of Cartesian coordinates remained restricted for all possible positions and orientations in respect to both pixels. Indeed, with the Cartesian co-ordinates, both pixels are only assumed have a common modal base when they are part of a single plane or two parallel planes, which is not necessarily the case every time. Therefore, when using the system of coordinates, it is difficult to simulate the coupling between any two pixels within free space.

Concerning the system of spherical coordinates both pixels may be integrated within the same plane. That is to say, it is an infinite beam sphere, if it is not a case of a straight perpendicular common beam, having defined norms with two pixels made up of the origins of either one or two spheres passing through two pixels. In the case of two pixels being part of two spheres, the latter may be non-concentric, which makes the determination of a modal base, which is common in these two pixels, difficult. In such a case, this system of coordinates is therefore not used.

By using cylindrical coordinates, both pixels may be located upon a single plane. This applies to a cylinder with an infinite beam, otherwise, they may be part of, either a single cylinder, or two coaxial cylinders. In reality, the straight perpendicular for both pixels constitutes the axis for the revolution of one cylinder or two coaxial cylinders passing through the two pixels (see Figure 6.43).

In summary, the cylindrical coordinate system is the only one which makes it possible to simulate any structure shape using the wave formulation.

Two pixels orientated and positioned in any way within the spaces can be part of either a single fictitious cylinder, or two fictitious coaxial cylinders. However, to determine the extent of these hypothetical cylinders, we can use the following technique:

Draw the normal for each pixel of the structure. The straight normal shown in relation to two normal versors constitutes the revolving axis of a single fictitious cylinder, provided that both pixels have the same radial distance (Figure 6.42). If not, this normal will be the axis of two coaxial cylinders. This situation is extremely well described by Figure 6.43.

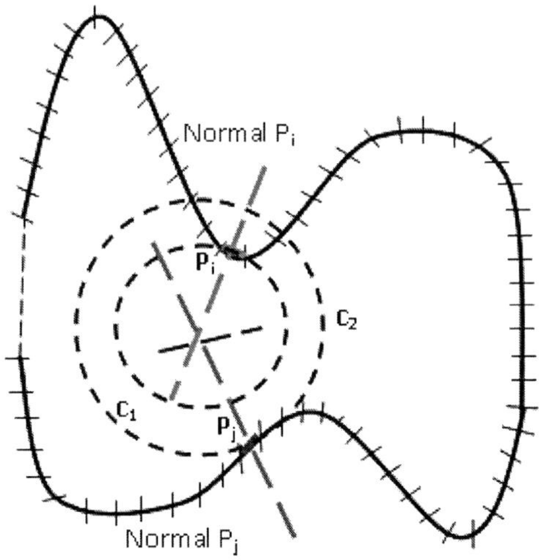

Figure 6.43. *Position of two pixels across two fictitious co-axial cylinders*

Both of these cases are likely to be easily replicated owing to modal diffraction operators described in the first section. In the case of a cylinder, external and internal diffraction operators in cylinder (C) occur within the replication process. In the case of two coaxial cylinders (C_1) and (C_2), three diffraction operators contribute to the modeling: the operator which is external to the cylinder (C_2), the operator which is internal to the cylinder (C_1) and that ensuring the transition between the cylinders (C_1) and (C_2). During the modeling phase, we should take account of both the different wave positions and directions.

6.3.1. *Different operators involved*

To determine the diffraction operators necessary for modeling the reaction of the environment around two pixels being placed in free space, we should take into consideration:

– the position of both pixels across fictitious cylinders, that is to say whether both pixels are part of a single fictitious cylinder or two coaxial fictitious cylinders;

– the positions of waves both emitted and received compared with the fictitious cylinder surfaces;

– the directions of waves which are both emitted and received compared with fictitious cylinder surfaces.

The modal diffraction operator is characterized by considering diffracted waves in free space as if the pixels did not exist (boundary conditions are not involved in the modeling).

6.3.2. *The case of two pixels on a single fictitious cylinder*

We now consider the case in which the two pixels are located upon the same fictitious cylinder; the various positions and orientations relative to waves both emitted $\vec{A}_i$ and received $\vec{B}_j$ (i and j represent the indices 1 and 2) are given in Figure 6.44.

In respect to these different coupling cases, modeling involves both internal and external diffraction operators upon cylinder (C) when in the modal domain. Figure 6.44 summarizes the various relevant wave positions and orientations in relation to cylinder (C).

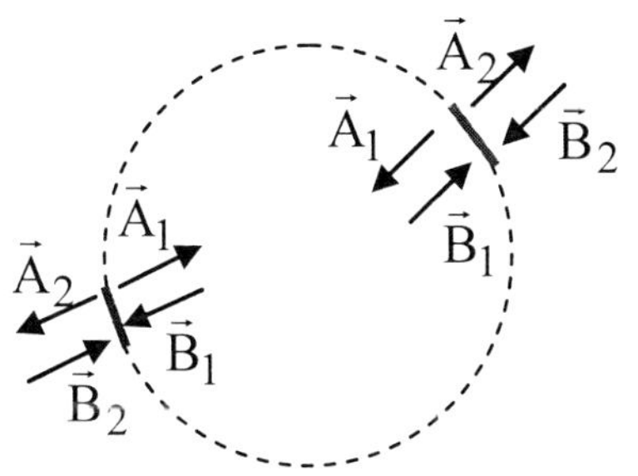

Figure 6.44. *Definition of two-pixel waves on a single cylinder*

The operator descriptions result from the equivalent diagram in Figure 6.45. They are governed by the following equations:

$$\begin{cases} \vec{B}_1 = \hat{\Gamma}_{in} \vec{A}_1 \\ \vec{B}_2 = \hat{\Gamma}_{ext} \vec{A}_2 \\ \vec{A}_1 = \vec{B}_2 \\ \vec{A}_2 = \vec{B}_1 \end{cases} \qquad [6.20]$$

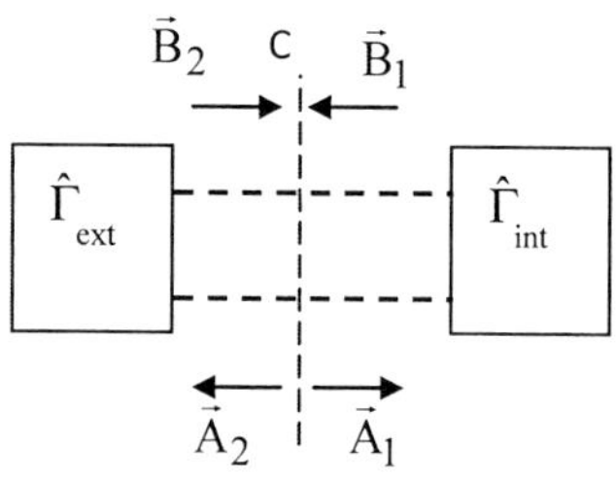

Figure 6.45. *Corresponding wave-based diagram*

From using the system of equations [6.20], for the various cases shown upon Figure 6.44, the waves ($\vec{B}_2$ and $\vec{B}_1$) are expressed according to ($\vec{A}_2$ and $\vec{A}_1$) by the modal diffraction operators which are a series of operators $\hat{\Gamma}_{ext}$ and $\hat{\Gamma}_{in}$. These operators are shown in Table 6.1.

Waves emitted	$\vec{A}_2$	$\vec{A}_2$	$\vec{A}_1$	$\vec{A}_1$
Waves received	$\vec{B}_2$		$\vec{B}_1$	$\vec{B}_2$
Operator	$\hat{\Gamma}_{ext}$	$\hat{\Gamma}_{ext}\ \hat{\Gamma}_{in}$	$\hat{\Gamma}_{in}$	$\hat{\Gamma}_{in}\ \hat{\Gamma}_{ext}$

Table 6.1. *The various diffraction operators in the case of a single fictitious cylinder*

For any two pixels within the space, the case in which the straight perpendicular in relation to their normals makes up the revolving axis of a single fictitious cylinder going through these two pixels, the representative diffraction operator for the interaction is determined by both the external and diffraction operators of cylinder (C).

6.3.3. *The case where the two pixels are part of two coaxial cylinders*

When the two pixels are located at different radial distances from the common perpendicular in relation to their normal versors, they are therefore part of two coaxial cylinders. The various positions and orientations of waves which are both emitted and received from the two pixels considered are shown in Figure 6.46. The modal diffraction operators which model the coupling between the two pixels are made up of the internal diffraction operator for the cylinder (C_1), the internal diffraction operator for the cylinder (C_2) and the transition operator which ensures wave transition between the cylinder (C_1) and the cylinder (C_2). It is also thus important to point out that fictitious cylinders are only involved in modeling so as to position those waves which are emitted and received in relation to each other.

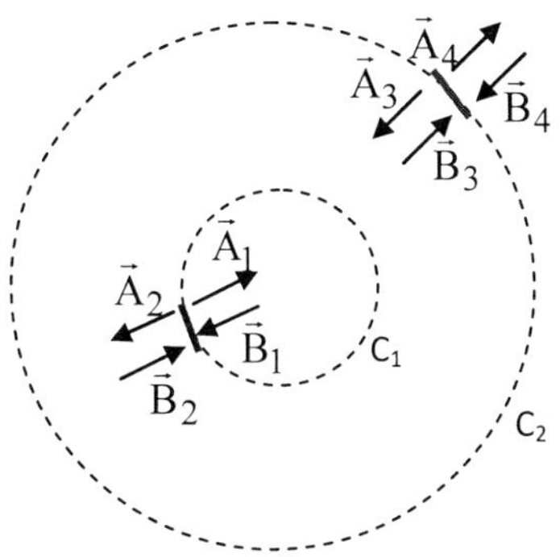

Figure 6.46. *Definition of two-pixel waves placed across two co-axial cylinders*

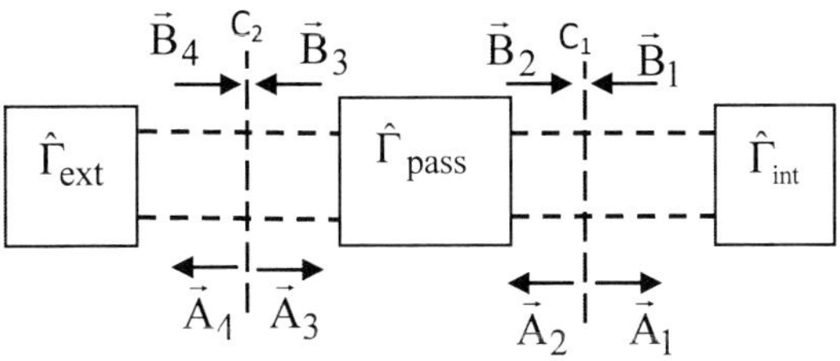

Figure 6.47. *Modeling two-pixel waves which are part of two co-axial cylinders*

Waves $\vec{B}_2$ received across the external surface of cylinder C_1 and $\vec{B}_3$ defined across the internal C_2 surface are connected to waves, denoted $\vec{A}_2$ which are emitted from the external surface of cylinder C_1 and denoted $\vec{A}_3$ from the interior surface of cylinder C_2 by using the equations:

$$\begin{cases} \vec{B}_2 = \hat{\Gamma}_{pass}^{11} \vec{A}_2 + \hat{\Gamma}_{pass}^{12} \vec{A}_3 \\ \vec{B}_3 = \hat{\Gamma}_{pass}^{21} \vec{A}_2 + \hat{\Gamma}_{pass}^{22} \vec{A}_3 \end{cases} \quad [6.21]$$

Taking the formula: $\hat{\Gamma}_{pass} = \begin{pmatrix} \hat{\Gamma}^{11}_{pass} & \hat{\Gamma}^{12}_{pass} \\ \hat{\Gamma}^{21}_{pass} & \hat{\Gamma}^{22}_{pass} \end{pmatrix}$

The equations which connect the various waves which are both received and emitted from one part or another of the two pixels are governed by the series of equations in [6.22]:

$$\begin{cases} \vec{B}_1 = \hat{\Gamma}_{in}\vec{A}_1 \\ \vec{B}_2 = \Gamma^{11}_{pass}\vec{A}_2 + \Gamma^{12}_{pass}\vec{A}_3 \\ \vec{B}_3 = \Gamma^{21}_{pass}\vec{A}_2 + \Gamma^{22}_{pass}\vec{A}_3 \\ \vec{B}_4 = \hat{\Gamma}_{ext}\vec{A}_4 \\ \vec{A}_1 = \vec{B}_2 \\ \vec{A}_2 = \vec{B}_1 \\ \vec{A}_3 = \vec{B}_4 \\ \vec{A}_4 = \vec{B}_3 \end{cases} \qquad [6.22]$$

From this system, the descriptors of modal diffraction operators may be determined by expressing waves received according to emitted waves under consideration.

Waves received	Waves emitted	Operator descriptors
$\vec{B}_4$	$\vec{A}_1$	$\left(II - \hat{\Gamma}_{ext}\hat{\Gamma}^{22}_{pass}\right)^{-1}\hat{\Gamma}_{ext}\hat{\Gamma}^{21}_{pass}\hat{\Gamma}$
$\vec{B}_1$	$\vec{A}_4$	
$\vec{B}_4$	$\vec{A}_2$	$\left(II - \hat{\Gamma}_{ext}\hat{\Gamma}^{22}_{pass}\right)^{-1}\hat{\Gamma}_{ext}\hat{\Gamma}^{21}_{pass}$
$\vec{B}_2$	$\vec{A}_4$	
		$\left(II - \hat{\Gamma}^{22}_{pass}\hat{\Gamma}_{ext}\right)^{-1}\hat{\Gamma}^{21}_{pass}\hat{\Gamma}_{in}$
$\vec{B}_1$	$\vec{A}_3$	
$\vec{B}_3$	$\vec{A}_2$	$\left(\mathrm{II} - \hat{\Gamma}^{22}_{\mathrm{pass}}\hat{\Gamma}_{\mathrm{ext}}\right)^{-1}\hat{\Gamma}^{21}_{\mathrm{pass}}$
$\vec{B}_2$	$\vec{A}_3$	

Table 6.2. *The various diffraction operators in the case of two coaxial cylinders*

When the pixels are at distances which are different from the perpendicular in relation to their normals, they are placed across two fictitious coaxial cylinders. The various possible cases of wave positions and directions induce modal diffraction operators. These diffraction operators are determined by diffraction operators which are external to the cylinder (C_2), the internal diffraction operator cylinder C_1 and the transition operators between two cylinders. Seven diffraction operators are determined for all of the possible positions and orientations of waves emitted and received. By defining these operators, WCIP is capable of modeling the coupling between two pixels which are both positioned and orientated in any given manner within free space.

6.3.4. *Spatial descriptors of diffraction operators*

For the 12 diffraction operators cited in the previous section, each was developed within the cylindrical modal base by the equation:

$$\hat{\Gamma}^{\sigma} = \sum_{n} \sum_{m} \left| f_{n,R_B}(\phi) \right\rangle \Gamma_n^{\sigma} \left\langle f_{n,R_A}(\phi') \right| \qquad [6.23]$$

with σ designating the component φ or z;

Γ_n^{σ} being the modal operator coefficients $\hat{\Gamma}^{\sigma}$;

R_A and R_B the cylinder beams where they are distinguished as emitted $\vec{A}$ and received waves $\vec{B}$.

We may recall that:

$f_{n,a}(\varphi) = \dfrac{1}{\sqrt{2\pi a}} e^{in\varphi}$ is the two polarizations of a modal base of a cylinder beam TE and TM.

The waves $\vec{A}$ and $\vec{B}$ are connected together within the spatial domain through equation [6.24]:

$$B^{\sigma}(\varphi) = \int_0^{2\pi} \Gamma^{\sigma,\tau}(\varphi,\varphi') A^{\tau}(\varphi') R_A d\varphi' \qquad [6.24]$$

$A(\phi')$ are waves emitted from the point ϕ';

$B(\phi)$ are waves received at the point ϕ;

$\Gamma^{\sigma}(\phi,\phi') = \sum_n \sum_m f_{n,R_B}(\phi)\Gamma_n^{\sigma} f_{n,R_A}^{*}(\phi')$ being the equation for the operator core $\hat{\Gamma}^{\sigma}$.

To characterize the various pixels of the structure, test functions $\{\left|\vec{g}_p\right\rangle\}$ are Heaviside step units defined so that the equation below applies:

$$g_p = \begin{cases} \dfrac{1}{\sqrt{\delta l}} & on\ a\ pixel.p \\ 0 & elsewhere \end{cases} \qquad [6.25]$$

The projection of [6.24] across the test functions in [6.25] gives the equation:

$$b_p^{\sigma} = \sum_{q=1}^{N} \Gamma_{p,q}^{\sigma} a_q^{\sigma} \qquad [6.26]$$

With the descriptors:

– b_p^{σ} and a_q^{σ} represent, respectively, the amplitudes of waves across the pixels p and q shown by the equation.

$$\Gamma^{\sigma}{}_{p,q} = \left\langle g_p \Gamma^{\sigma} g_q \right\rangle = \sum_{n=-\infty}^{n=-\infty} \left\langle g_p \middle| f_{n.R_B}(\varphi) \right\rangle \Gamma^{\sigma}{}_n \left\langle f_{n.R_A}(\varphi') \middle| g_q \right\rangle \qquad [6.27]$$

By completing the scalar product, which is defined in [6.27], the analytical descriptor of the diffraction operator will be the following equation:

$$\Gamma^{\sigma}(p,q) = \frac{\delta l}{2\pi\sqrt{R_p R_q}} \sum_{n=-\infty}^{n=-\infty} \Gamma^{\sigma}{}_n e^{j*n\left(\frac{p}{R_p} - -\frac{q}{R_q}\right)\delta l} \sin c^2\left(\frac{n\delta l}{2}\right) \qquad [6.28]$$

In [6.28], the analytical descriptor of $\Gamma^{\sigma}(p,q)$ represents the coupling between two pixels (p and q), placed in any way whatsoever within the relevant space.

6.3.5. *The iterative process*

The iterative process is governed by a two-equation system, entirely defined within the spatial domain [6.29]. One equation expresses the continuity equations of electromagnetic fields in terms of wave amplitudes, across each discontinuity interface, in the presence of an excitation source. The second equation connects wave amplitudes of incident waves to reflected waves through the intermediary of spatial diffraction operators:

$$\begin{cases} a^{\sigma} = \hat{S}\, b^{\sigma} + (a_0^{\sigma}, b_0^{\sigma}) \\ b^{\sigma} = \Gamma^{\sigma} b^{\sigma} \end{cases} \qquad [6.29]$$

Using the equations:

$$b^{\sigma} = (b_2^{\sigma}, b_1^{\sigma}),\ a^{\sigma} = (a_2^{\sigma}, a_1^{\sigma}) \text{ and } \Gamma^{\sigma} = \begin{pmatrix} \Gamma_2^{\sigma} & 0 \\ 0 & \Gamma_1^{\sigma} \end{pmatrix}$$

upon the convergence of the iterative process, waves emitted and received are determined by their amplitudes. Next, the electric field and the current density are calculated, by using wave definition relationships.

6.3.6. *Computation of the remote location electric field*

Upon the convergence of the iterative process, once the waves $\vec{A}$ and $\vec{B}$ are calculated across the surface of the structure (Figure 6.48), the same technique, used before, will be applied to calculate the infinite magnetic fields. In fact, the $\vec{A}$ waves emitted from the external surface of the structure will be brought across a fictitious cylinder with an infinite beam ($\rho >> \lambda$) to create waves $\vec{B}^{\text{inf}}$ (Figure 6.48). The intersection of the beam (ρ) of the infinite cylinder and the normal in relation to a pixel of the structure constitutes the center of coaxial virtual cylinders. Taking account of the asymptotic development of the Bessel and Hankel functions, the operators defined within the first section are applied to waves $\vec{A}$ to calculate infinite $\vec{B}^{inf}$ waves.

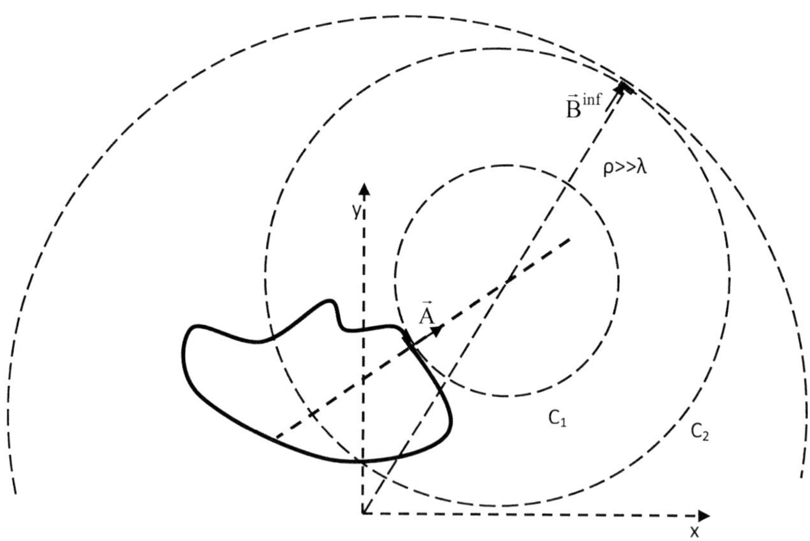

Figure 6.48. *Calculation of diffracted fields in the remote zone*

6.3.7. *Application*

To validate the spatial approach of the WCIP method, some examples of applications, which take account of diffraction issues, will now be explained.

6.3.7.1. *Right-angled metallic corner*

Figure 6.49 shows a right-angled metallic corner placed in free space and which rests upon an absorbent media (see [AMM 06, AGU 06, BAU 06]). Each edge of the corner assumes a length $l = 4*\lambda$ and is discretized in $\lambda/40$. The structure is excited by a plane wave TM such that the equation below applies:

$$\vec{E}^{\,incid} = E_0 e^{jk_0\left(-x\cos(\varphi_0)+y\cos(\varphi_0)\right)}\vec{z} \qquad [6.30]$$

using the equation $E_0 = 1$.

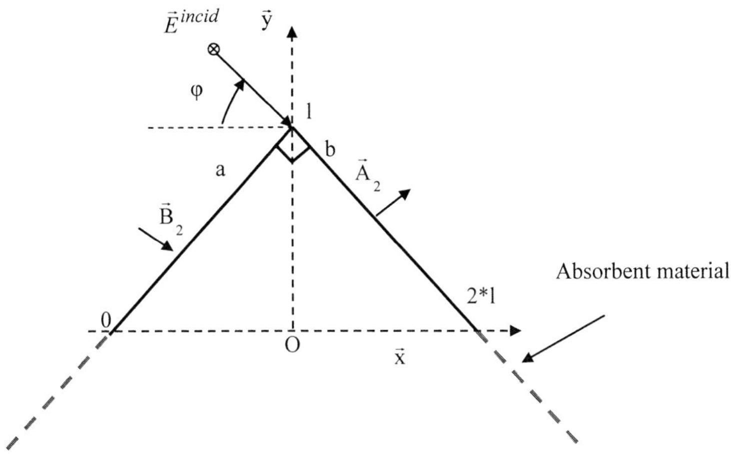

Figure 6.49. *Right-angled metallic corner resting upon absorbent material*

The total current density is given by:

$$J_z = \frac{1}{\sqrt{Z_0}}(A_2 - B_2) + J^{incid} \qquad [6.31]$$

J^{incid} is the current density associated with the incident field.

The simulation results of the current density are given for the incident-field equations $\varphi_0 = \pi/4$ and $\varphi_0 = \pi/2$.

In the case of $\varphi_0 = \pi/4$ the incident wave is perpendicular to edge (a) of the metallic corner, therefore it does not meet side (b) (the covered area) which leads to the induced current being negligible across line (b) which is edge (a) (Figure 6.50). For $\varphi_0 = \pi/2$ the distribution of the incident field is symmetrical across both edges of the corner which leads to a current density which is symmetrical (Figure 6.51)

After convergence, the waves $\vec{A}_2$, emitted from edges (a) and (b), are transformed to infinity to calculate the diffracted field in the remote zone. Figures 6.52 and 6.53 show the *z* components of the modulus of the diffracted electric fields within the remote location zone in respect to angles of incidence $\varphi_0 = \pi/4$ and $\varphi_0 = \pi/2$. These curves show that for an incidence wave $\varphi_0 = \pi/4$ the field which is diffracted is at its maximum for $\varphi = \pi/4$. On the other hand, for an incidence wave $\varphi_0 = \pi/2$ the electric field shows two maximum values in $\varphi = 0$ and $\varphi = \pi$.

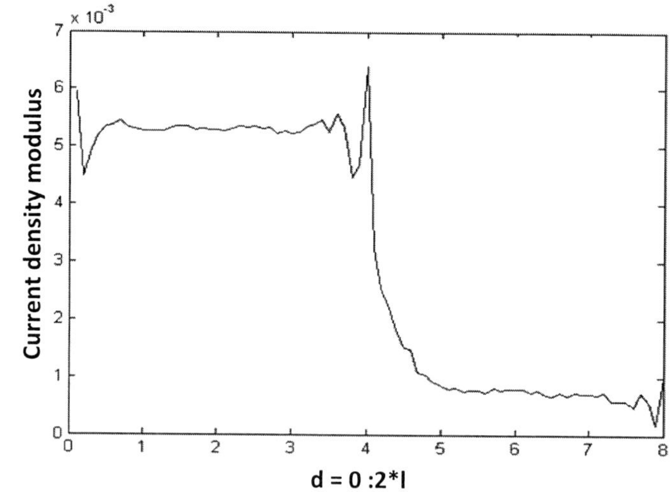

Figure 6.50. *Current density for the equation $\varphi_0 = \pi/4$*

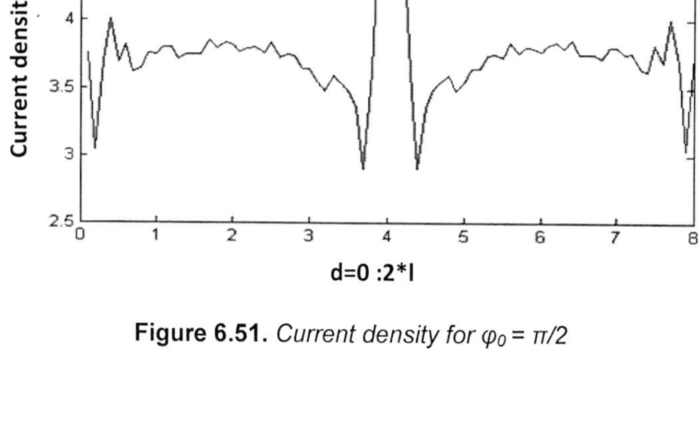

Figure 6.51. *Current density for $\varphi_0 = \pi/2$*

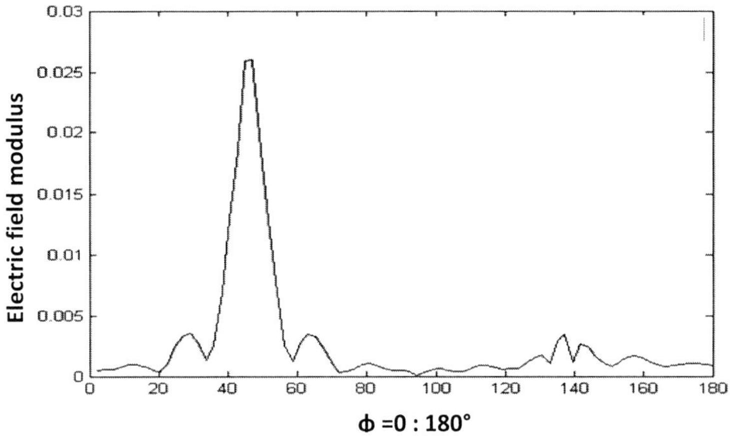

Figure 6.52. *Electric field in the remote zone for $\varphi_0 = \pi/4$*

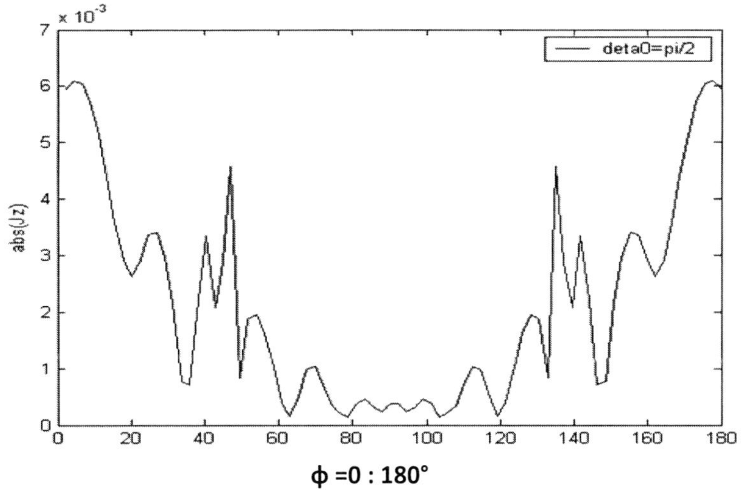

Figure 6.53. *Electric field in the remote zone for $\varphi_0 = \pi/2$*

6.3.7.2. *Metallic corner of the angle α = 120°*

In this example, a metallic corner with an aperture $\alpha = 120°$ is considered. Each side is made up of a metallic part with a length $l_m = 5*\lambda$ and a dielectric part of a length $l_d = 4*\lambda$. The structure is meshed with a mesh density equal to $\lambda/20$. The corner is excited by a linear current source, which is placed at a distance $d = 2*\lambda$ in relation to the apex of the corner (Figure 6.54). The electric field generated by this source is given as the formula:

$$\vec{E}_z^{incid} = -\frac{k_0 z_0}{4} H_0^{(2)}(k_0 \rho_m)\vec{e}_z \qquad [6.32]$$

Using the symbol:

ρ_m, which is the distance between the source and each pixel of the structure.

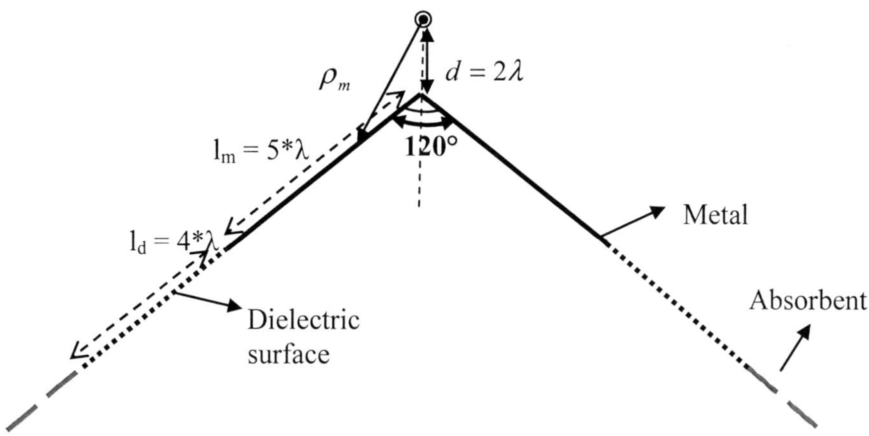

Figure 6.54 *Metallic corner with an aperture of α = 120°*

Current density shown across both metallic sides is compared with that given by the method of moments (see [SHI 05, AND 05]).The comparison of these result reveals a perfect resemblance between the methods used (Figure 6.55).

6.3.7.3. *Diffraction by a square cylinder of an infinite length*

Figure 6.56 shows a square-section cylinder, of an infinite length in the direction z, the length of one edge is $a = 8*\lambda$ (see [AMM 06 AGU 06, BAU 06]). Each side discretized with a mesh density equal to $\lambda/20$.

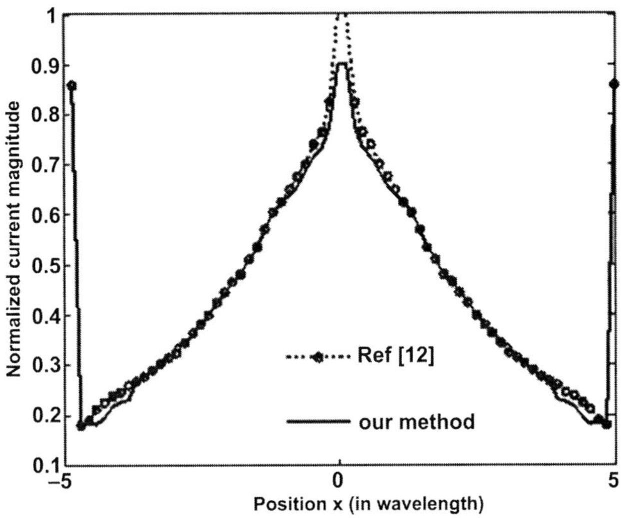

Figure 6.55. *Standardized current density*

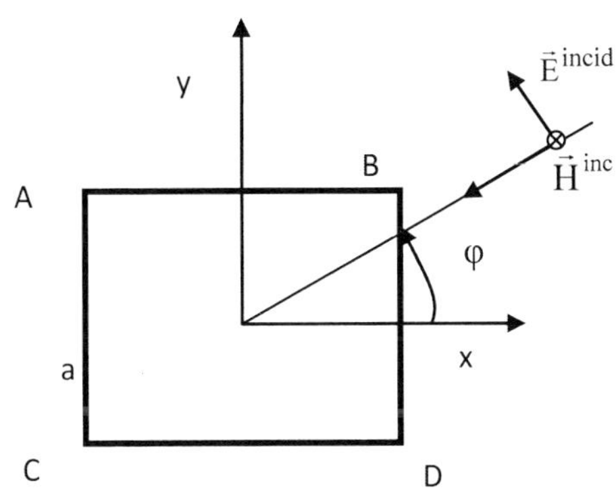

Figure 6.56. *Transverse section of a square cylinder*

The structure is excited by a plane wave TE, such that the incident magnetic field is shown by the equation:

$$\vec{H}^{incid}(x,y)=H_0 e^{-jk_0(x\cos(\varphi_0)+y\sin(\varphi_0))}\vec{u}_z \qquad [6.33]$$

Using the equation $H_0 = 1$.

Figure 6.57 shows the variation in current density across the two edges (shaded side [CD] and the exposed side [BD]) for an angle of incidence with the formula $\varphi_0 = \pi/4$. The result given by the iterative method is compared with that given by the method of moments (as seen in [PAN 03, TRE 03, GIL 03]). Despite these minor errors which are essentially due to the imprecision of the mesh, we can say that there is a close alignment between the two methods used.

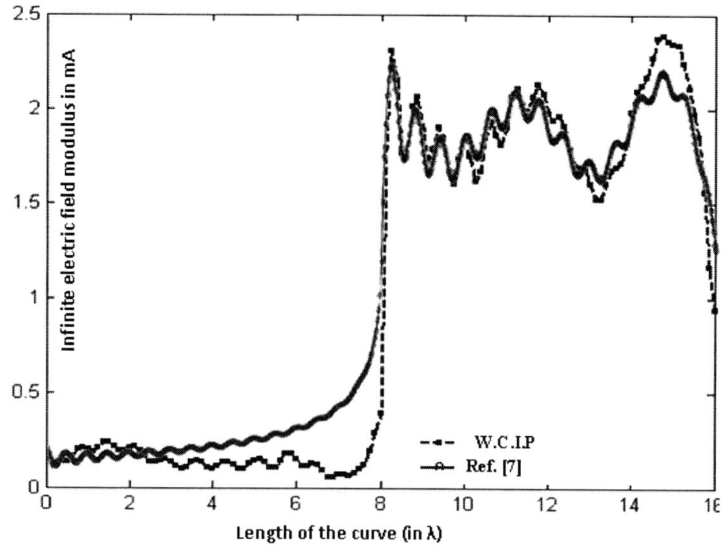

Figure 6.57. *Standardized current density*

The diffracted electric field within a remote location zone is shown for $\varphi \in [-45, 135°]$ (Figure 6.58). The variation of the modulus of the standardized diffracted field shows an initial maximum for ($\varphi = -45^0$ and $\varphi = 135^0$) and second maximum for ($\varphi = 0°$ and $\varphi = 90°$). These maxima are validated by the geometric perspective. Indeed, the first premier maximum is an accumulation of the diffracted fields and peak effects, the second maximum is explained by the geometric angles of diffracted fields.

6.3.7.4. *Shaped diffraction in L*

Within this application, a structure is shown and studied in the geometric form of L ((Figure 6.59) [AMM 08]). It is considered infinite in the z direction, the length of the large edge is $l = 6.4*\lambda$. The structure is excited by a wave plane TM, with an angle of incidence of $\varphi = 45°$.

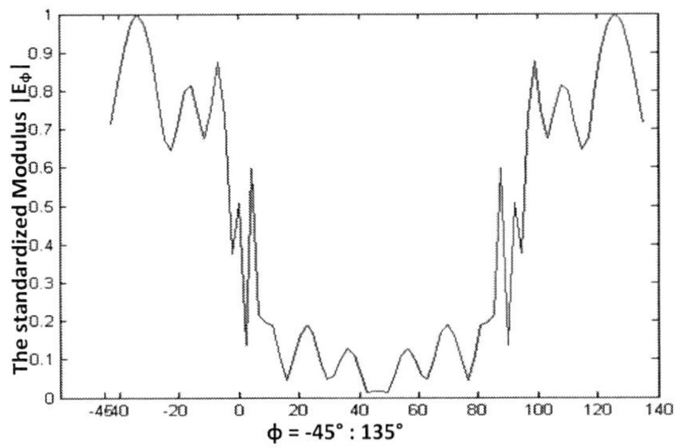

Figure 6.58. *Variation of the electric field within the remote zone*

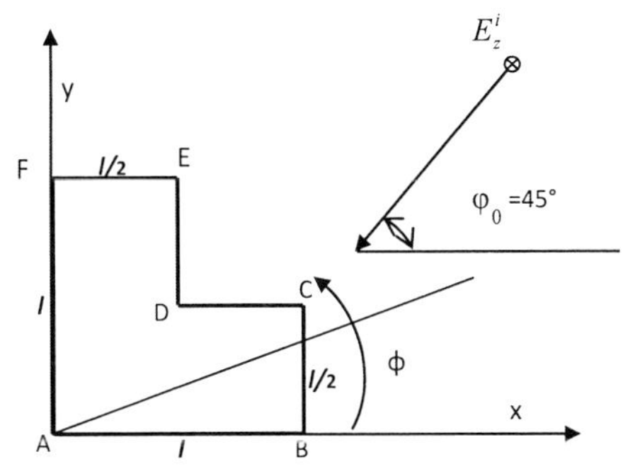

Figure 6.59. *Transverse section of the structure*

Considering the symmetry of the structure, the current density is shown by the curve (AD) (Figure 6.60). The result of the simulation shows a very strong similarity between the WCIP method and the reference (see [PAN 03, TRE 03, GIL 03]).

The analysis of curves given in Figure 6.61, for an angle of incidence $\varphi_0 = 45°$, shows that the echo radar, standardized in relation to wave length, shows a principal maximum at $\varphi = 45°$, and the secondary maximum at the following points$\{\varphi = 135°, \varphi = 225°$ and $\varphi = 335°\}$. At the point $\varphi = 45°$ the value of the diffracted electric field in the remote zone results from the fields diffracted by points (A, B, C, and F) of the structure. By relying upon the geometric optics, the secondary maxima correspond to the diffraction angled by the various edges of the structures. It turns out that there is thus a very close alignment between the solution provided by the WCIP method and that obtained by the reference study (see [PAN 03, TRE 03, GIL 03]).

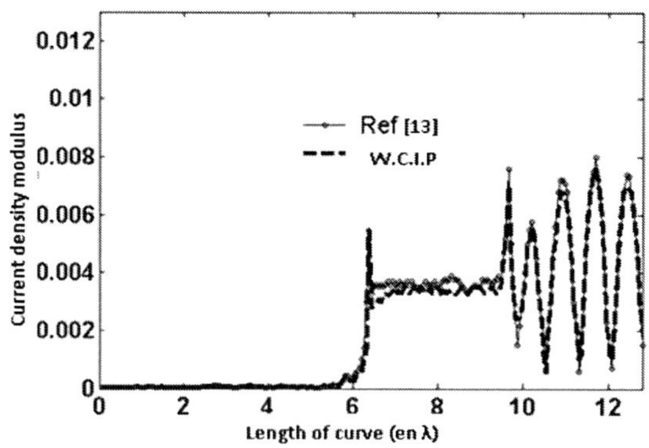

Figure 6.60. *Distribution of current density*

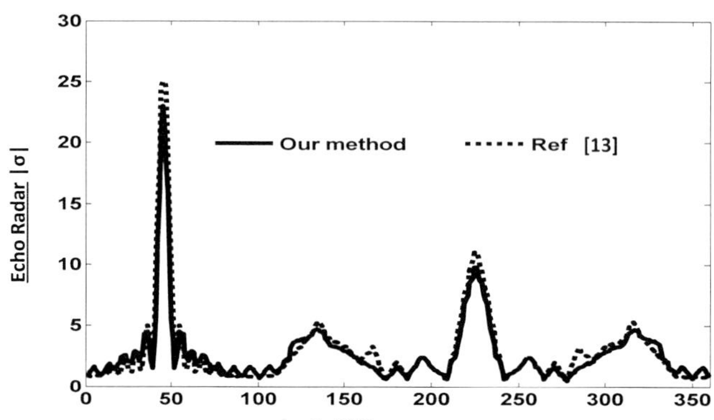

Figure 6.61. *Diffraction coefficient in the standardized remote zone*

6.3.7.5. *Photonic or electromagnetic beam*

The photonic beam is a narrow high-intensity light beam, which emerges from the covered area of a cylinder or from a dielectric sphere, which demonstrates no losses, with a beam that is comparable to its wavelength, illuminated by a wave plane. This phenomenon was discovered by [TAF 04] (see [CHE 04 TAF 04]) upon lighting a dielectric cylinder of an infinite length across a wave plane.

Some key properties of the photonic jet are:

1) it has an increased intensity compared with the light wave;

2) the maximum mid-height width of the beam is less than the wave length;

3) the beam propagates with a lower divergence for several wavelengths.

The phenomenon of the photonic jet is examined in the microwave domain and certain measurements are made in the Ka band. We have chosen to use the general term of "electromagnetic" jet to describe the phenomenon.

Figure 6.62 explains the geometry of the structure which generates the photonic jet [IYE 03]. It consists of a bi-plate waveguide integrated with teflon (of permittivity $\varepsilon_r = 2$) the structure has a height of 3 cm and a width which is theoretically considered to be infinite (but here measuring 20 cm). The guide is closed by an elliptical Teflon tip of the semi-major axis $b = 4$ cm and semi-minor axis $a = 1.5$ cm. Both electric walls are restricted at their extremities by two infinite metallic plates. The guide is excited by the electric transverse fundamental mode.

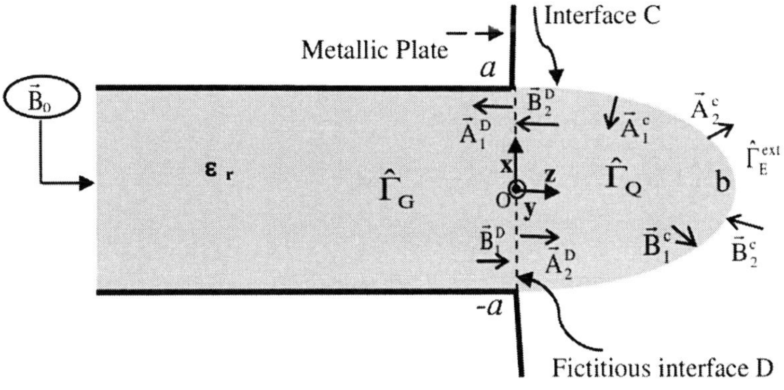

Figure 6.62. *Structure which generates an electromagnetic jet [AMM 15]*

The iterative process for this new approach for the WCIP method is governed by two systems of equations which are entirely defined within the spatial domain. The first system expresses continuity equations within electromagnetic fields in terms of waves across discontinuity interfaces D and C (Figure 6.63) where these waves are linked by the observed operators: $\hat{S}_D$ and $\hat{S}_C$. The system of equations used expresses the reactions of spaces surrounding the discontinuity interfaces, as well as the coupling between these two interfaces. The iterative process is activated by the excitation source. Next, the device switches between both equation systems until convergence occurs.

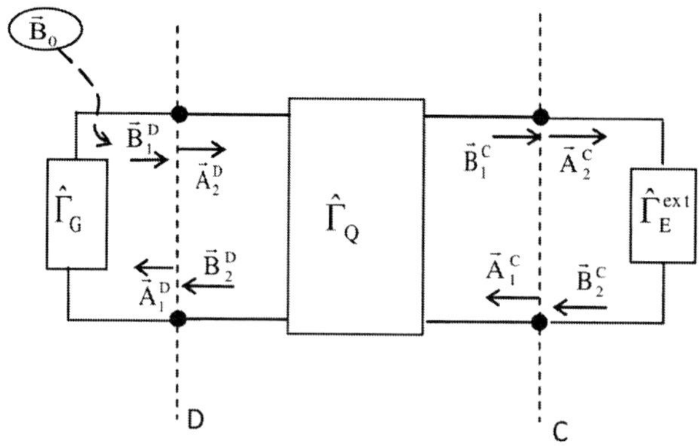

Figure 6.63. *Equivalent diagram in waves*

$\vec{B}_0$ represents the power source in terms of waves and the equation is expressed as

$$\vec{B}_0(x) = \frac{E_0}{\sqrt{z_{0.1}}}\sin(\frac{\pi x}{2a})\vec{y} \qquad [6.34]$$

As E_0 represents the source amplitude, its value is equal to 1.

In this particular application, we are concerned with the computation of the near-field opposite the elliptical tip. To compute these fields, after the convergence within the iterative process, waves $\vec{A}_2^c$ emitted from the exterior of the interface C are sent back to the given observation points. To do this, the same modeling technique used previously is applied. In practice, these observation points are considered as pixels placed along a fictitious line denoted DF (Figure 6.64) with the equation:

$$z = b + z' \qquad [6.35]$$

Using z' ≥ 0

The waves $\vec{B}_{1.q}^{E}$ received at pixel (q) from the fictitious interface D_F are established from the emitted wave $\vec{A}_{2.p}^{c}$. By varying z', we are able to deduce the values of fields at any point within the limited half-space by the equation $z = b$.

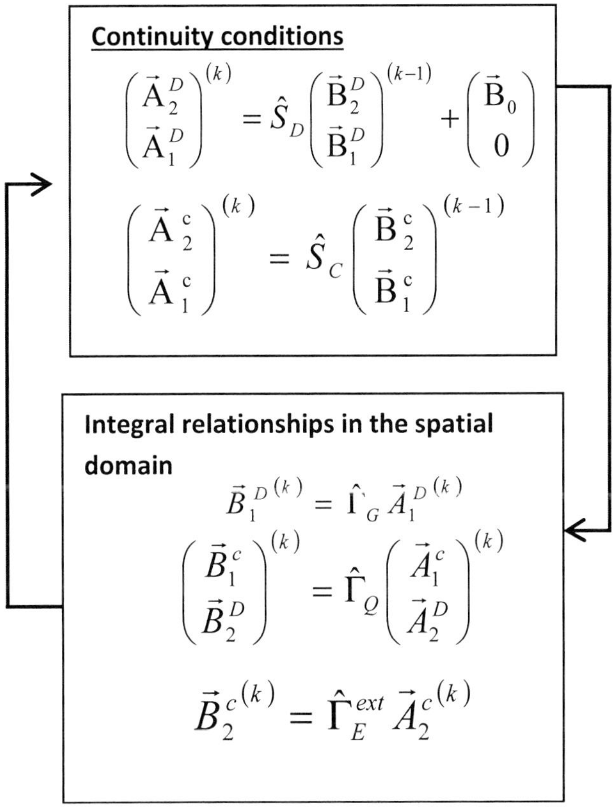

Figure 6.64. *Diagram of the iterative process*

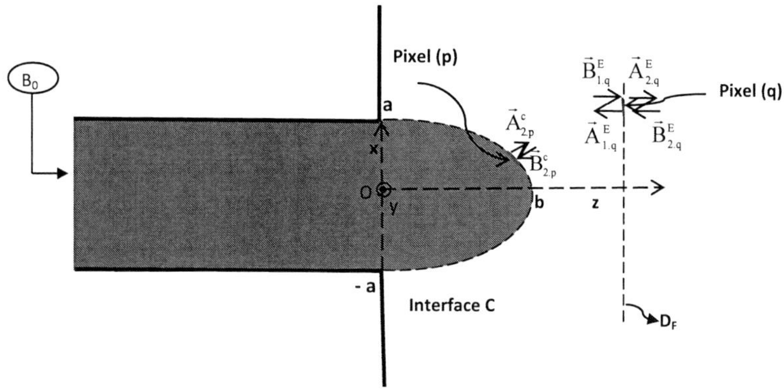

Figure 6.65 *Computation of the near field*

Figures 6.66(a) and (b) show variations in electric fields across the interface C for two power source frequencies; f = 30 GHz and f = 60 GHz. This corresponds respectively to wave lengths within the dielectric surface of λ_d = 7.1 mm and λ_d = 3.6 mm. Theoretically, for f = 30 GHz, 25 resonance frequency pikes are spread along the 9 cm interface C. The distance between two successive peaks is equal to $\lambda d/2$ or 3.6 mm. This simulation is approved by the WCIP method within Figure 6.66(a). Upon doubling the excitation frequency (f = 60 GHz), Figure 6.66(a) further shows 50 resonance frequencies where the distance between two successive peaks is equal to 1.8 mm.

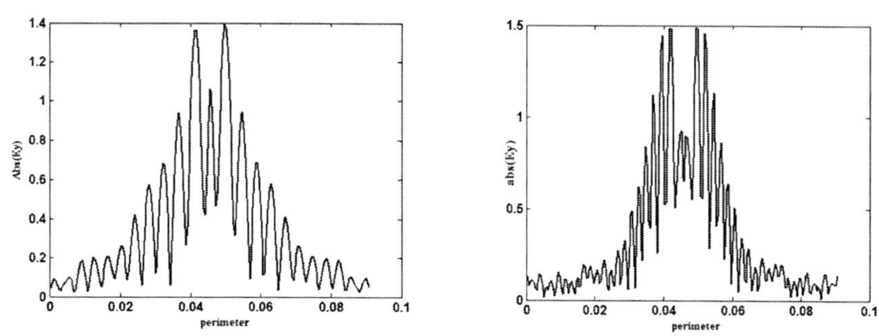

(a) f = 30 Ghz (b) f = 60 Ghz

Figure 6.66. *Variation of the electric field across the interface C (see [AMM 15])*

Figure 6.67 shows the distribution of the normalized electric field opposite the elliptical peak at 30 GHz, for various values of relative permittivity. The structure dimensions are a = 1.5 cm and b = 4 cm. It may be observed that the focal point moves towards the dielectric peak when permittivity increases (it will thus be within the elliptical peak), as well as the propagation distance diminishing.

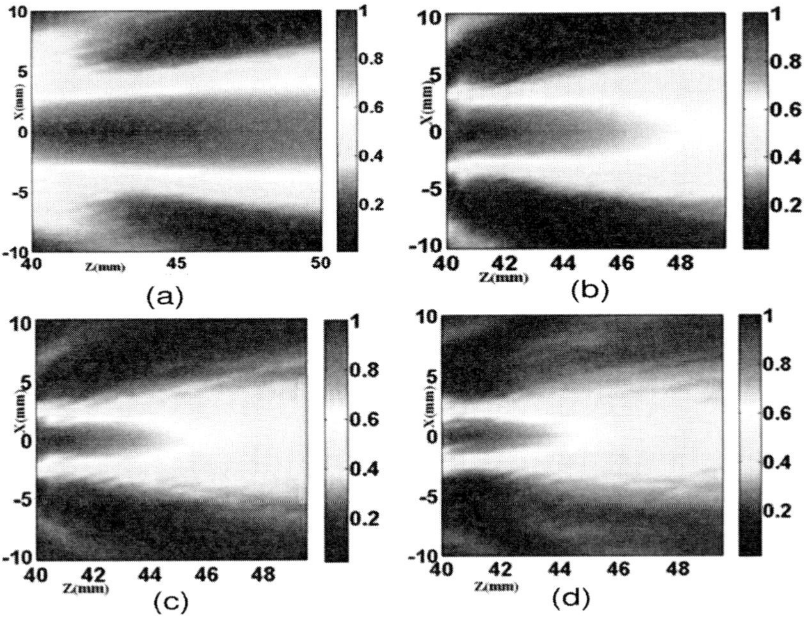

Figure 6.67. *Electric field modulus for various ε_r values: a) ε_r = 2, b) ε_r = 4, c) ε_r = 6 and d) ε_r = 8. For a color version of this figure, see www.iste.co.uk/baudrand/waves.zip*

6.4. Conclusion

In the first section of this chapter, wave formulation was applied to various inhomogeneous structures by using three types of diffraction operators which model the different sub-space reactions around the interfaces in a given case study. The structures studied in the different paragraphs have permitted us to show both the efficiency and the capacity of the WCIP method when distinguishing complex cylindrical structures. Within multilayer technology, the results obtained for the various devices analyzed are very close to those found within works in the field.

In the second section, the wave formulation as cylindrical coordinates was extended to the study of structures, which may take any shape, when placed within free space. Owing to the notion of fictitious cylinders, the coupling between two pixels, both of which are positioned and orientated in any way within free space, was distinguished by diffraction operators developed within cylindrical modal bases. Taking into consideration all of the positions and orientations possible for emitted and received waves at the level of each pixel pair, twelve diffraction operators specified within the modal domain are developed. Finally, these various modal operators were projected across the test functions to assume their given spatial shape.

For invariant structures in the *z* direction, various applications proved the performance qualities of this simulation tool. Indeed, this new approach successfully showed the expansion of cylindrical coordinate wave formulation through studying planar structures. The application to diffraction issues, through the use of geometrical structures made up of multiple metal corners, demonstrated a very close alignment between both the simulation results of current densities and the diffracted fields in remote zone locations, and given control studies within the field.

Bibliography

[ABD 01] ABDEL M.A., MONEUM SHEN Z. *et al.*, "Hybrid PO-MOM analysis of large axi-symmetrie radomes", *IEEE Transactions on Antennas and Propagation*, vol. 49, no. 12 pp. 1657–1666, 2001

[AKA 01] AKATIMAGOOL S., BAJON D., BAUDRAND H., "Analysis of multi-layer integrated inductors with wave concept iterative procedure (WCIP)", *IEEE MTT-S*, vol. 3, pp. 1941–1944, 2001.

[ALI 98] ALIMENTI F., TIEZZI F., SORRENTINO R. *et al.*, "Accurate analysis and modeling of slot coupled patch antennas by the TLM-IE and the FDTD methods", *28th European Microwave Conference*, vol. 1, pp. 30–35, October 1998.

[AMM 06a] AMMAR N., AGUILI T., BAUDRAND H., "Study of electromagnetic waves diffraction by rectangular cylinders using the "Wave concept iterative process" (WCIP) in spatial domain", *Mediterranean Microwave Symposium*, Genova, Italy, September 19–21, 2006

[AMM 06b] AMMAR N., AGUILI T., BAUDRAND H., "Study of electromagnetic waves diffraction by a right wedge using the "Wave Concept Iterative Process" (WCIP) in spatial domain", *IEEE International Conference on Information & Communication Technologies from Theory to Applications*, Damascus, Syria, April 24–28, 2006.

[AMM 08] AMMAR N., AGUILI T., BAUDRAND H., "Analysis of multilayered cylindrical structures using a full wave method", *Progress in Electromag. Research*, vol. 85, pp. 425–438, 2008.

[AMM 08] AMMAR N., TAREK B., AGUILI T. *et al.*, "Investigation of electromagnetic scattering by arbitrarily shaped structures using the wave concept iterative process", *Journal of Microwaves, Optoelectronics and Electromagnetic Applications*, vol. 7, no. 1, pp. 26–43, 2008.

[AMM 09] AMMAR N., AGUILI T., BAUDRAND H., "Modeling of wave penetration through multilayered cylindrical aperture using an iterative method based on transverse wave concept", *International Journal of Microwave and Optical Technology*, vol. 4, no. 1, pp. 27–34, 2009.

[AMM 15] AMMAR N., AGUILI T., BAUDRAND H. *et al.*, "Wave concept iterative process method for electromagnetic or photonic jets: numerical and experimental results", *IEEE Transactions Antennas Propagation*, vol. 63, no. 11, pp. 4857–4867, 2015.

[ANA 96] ANASTASSIU H.T., VOLAKIS J.L., ROSS D.C. *et al.*, "Electromagnetic scattering from Simple Jet Engine models", *IEEE Transactions on Antennas and Propagation*, vol. 44, no. 3, pp. 420–431, 1996.

[ARN 08] ARNIERI E., AMENDOLA G., "Analysis of substrate integrated waveguide structures based on the parallel-plate waveguide green's function", *IEEE Transactions on Microwave Theory and Techniques*, vol. 56, pp. 1615–1623 2008.

[AZI 13] AZIZI M. K., LATRACH L., RAVEU N. *et al*, "A new approach of almost periodic lumped elements circuits by an iterative method using auxiliary sources", *American Journal of Applied Sciences*, vol. 10, no.11, pp. 1457–1472, 2013.

[AZI 95] AZIZI M., AUBERT H., BAUDRAND H., "A new iterative method for scattering problems", *Proceedings of European Microwave Conference*, vol. 1, pp. 256–258, 1995.

[AZI 96] AZIZI M., BOUSSOUIS M., AUBERT H. *et al.*, "A three dimensional analysis of planar discontinuities by an iterative method", *Microwave and Optical Technology Letters*, vol. 13, pp. 372, 1996.

[BAU 09] BAUDRAND H., WANE S., "Multiscale circuits: use of auxiliary sources", in: SAGUET P. (ed.), *Passive RF Integrated Circuits: Modeling, Characterization and Measurement*, ISTE Ltd, London and John Wiley & Sons, New York, 2009.

[BAU 99] BAUDRAND H., GONGO R.S.N., "Application of wave concept iterative procedure in planar circuits", *Recent Research Development Microwave Theory and Technics,* Transword Research Network, vol. 1, pp.187–197, 1999.

[BOZ 01] BOZZI M., PERREGRINI L., WEINZIERL J. *et al.*, "Efficient analysis of quasi-optical filters by a hybrid MoM/BI-RME method", *IEEE Transactions Antennas Propagation*, vol. 49, no. 7, pp. 1054–1064, July 2001.

[CAM 02] CAMPOS A.L.P.S., D'ASSUNÇÃO A.G., GOMES NETO A., "Scattering characteristics of FSS on two anisotropic layers for incident co-polarized plane waves", *Microwave and Optical Technology Letters*, vol. 33, no. 1, pp. 57–61, 2002.

[CAS 02] CASSIVI Y., DESLANDES D., WU K., "Substrate integrated waveguide directional couplers", *Proceedings of Asia-Pacific Microwave Conference, Kyoto*, November 2002.

[CAS 04a] CASSIVI K., WU K., "Substrate integrated NRD (SINRD) guide in high dielectric constant substrate for millimeter wave circuits and system", *Microwave Symposium Digest, IEEE MTT-S International*, vol. 3, pp. 1639–1642, 2004.

[CAS 04b] CASSIVI Y., WU K., "Substrate integrated nonradiative dielectric Waveguide", *IEEE Microwave and Wireless Components Letters*, vol. 14, no. 3, pp. 89–91, 2004.

[CHE 04] CHEN Z., TAFLOVE A., "Photonic nanojet enhancement of backscattering of light by nanoparticles: a potential visible-light ultra microscopy technique", *Optics Express*, vol. 12, pp. 1214–1220, 2004.

[CHE 08a] CHEN X.P., WU K., "Substrate integrated waveguide cross coupled filter with negative coupling structure", *IEEE Transactions Microwave Theory Technology*, vol. 56, no. 1, pp. 142–149, 2008.

[CHE 08b] CHENG Y.J., HONG W., WU K., "Design of a monopulse antenna using a dual V-type linearly tapered slot antenna (DVLTSA)", *IEEE Transactions Antennas and Propagation*, vol. 56, no. 9, pp. 2903–2909, 2008.

[ERD 02] ERDEMLI Y.E., SERTEL K., GILBERT R.A. *et al.*, "Frequency-selective surfaces to enhance performance of broad-band reconfigurable arrays", *IEEE Transactions on Antennas and Propagation,* vol. 50, no. 12, pp. 1716–1724, 2002.

[FIC 08] FICHTNER N., WANE S., BAJON D. *et al.*, "Interfacing the TLM and the TWF method using a diakoptics approach", in *IEEE MTT-S International Microwave Symposium Digest,* pp 57–60, 2008.

[FIL 00] FILHO O.M.C.P., SARKAR T.K., "Full-wave analysis of MICs in multilayer dielectric media in a rectangular waveguide", *IEEE Transactions on Microwave Theory and Techniques,* vol. 48, no. 10, pp. 1611–1622, 2000.

[GAO 05] GAO Q., YAN D.B., YUAN N.C. *et al.*, "Resonance of compact frequency selective surface arrays", *Progress in Electromagnetics Research Symposium*, Hangzhou, pp. 23–26, 2005.

[GAR 01] GARCIA R., *Contribution à l'Etude de Circuits Planaires par une Méthode Itérative Basée sur le Concept d'Onde (F.W.C.I.P.)*, PhD Thesis, Laboratoire d'électronique de ENS, France, 2001

[GON 99] GONGO R.S.N., BAUDRAND H., "Application of wave concept iterative procedure in planar circuits", *Recent Research Developments in Microwave Theory and Techniques*, vol. 1, pp. 187–197, 1999

[GRB 04] GRBIC A., ELEFTHERIADES G.V., "Overcoming the diffraction limit with a planarleft-handed transmission-line lens," *Physical Review Letters*, vol. 92, no. 11, pp. 117–403, 2004.

[HAO 05] HAO Z.C., HONG W., CHEN J.X. *et al.*, "Compact super-wide bandpass substrate integrated waveguide (SIW) filters", *IEEE Transactions on Microwave Theory and Techniques*, vol. 53, no. 9, pp. 2968–2977, 2005.

[HAO 05] HAO Z.C., HONG W., CHEN J.X. *et al.*, "Planar diplexer for microwave integrated circuits", *IEEE Proceedings Microwave Antennas and Propagation*, vol. 152, no. 6, pp. 455–459, 2005.

[HAR 13] HAROUNI Z., OSMAN L., LATRACH L. *et al.*, "A novel analysis of a rectenna circuit with iterative method", *10th International Multi-Conference on Systems, Signals & Devices (SSD)*, pp. 1,4, March 2013.

[HIR 98] HIROKAWA J., ANDO M., "Single-layer feed waveguide consisting of posts for plane TEM wave excitation in parallel plates", *IEEE Transactions Antennas and Propagation*, vol. 46, no. 5, pp. 625–630, May 1998.

[HOM 80] HORNO M., "Calculation of quasi-static characteristics of microstrip on anisotropic substrate using mapping method", *Microwave Symposium Digest, MTT-S International*, vol. 80, no. 1, pp. 450–452, 1980.

[HON 06] HONG W., HAO Z.-C., LI H. *et al.*, "Development of a low cost microwave mixer using a broad-band Substrate Integrated Waveguide (SIW) coupler", *IEEE Microwave and Wireless Components Letters*, vol. 16, no. 2, pp. 84–86, 2006.

[HOR 80] HORNO M., "Calculation of quasi-static characteristics of microstrip on anisotropic substrate using mapping method," *Microwave Symposium Digest, MTT-S International*, vol. 80, no. 1, pp. 450–452, 1980.

[IOA 05] IOANNIS I. HERETAKIS, PANAGIOTIS J. *et al.*, "A stochastically optimized adaptive procedure for the location of MAS auxiliary monopoles: the case of electromagnetic scattering by dielectric cylinders", *IEEE Transactions on Antennas and Propagation,* vol. 53, no. 3, pp 938–947, 2005.

[ISI 13] ISIDRO R., COVES A., SANCHEZ-SORIANO M.A. *et al.*, "Systematic study of the effective permittivity in a periodically drilled substrate integrated waveguide", *PIERS*, Stockholm, 12–15 August 2013.

[ISM 11] ISMAIL ALHZZOURY A., RAVEU N., PRIGENT G. *et al.*, "Substrate integrated waveguide filter design with wave concept iterative procedure", *Microwave and Optical Technology Letters*, vol. 53, no. 12, pp. 2939–2942, 2011.

[ISM 12] ISMAIL ALHZZOURY A., RAVEU N., PIGAGLIO O. *et al.*, "WCIP applied to substrate integrated waveguide", *PIER C*, vol. 33, pp 171–184, 2012.

[ISM 14] ISMAIL ALHZZOURY A., RAVEU N., PIGAGLIO O. *et al.*, "SINRD circuits analysis with WCIP", *PIER C*, vol. 52, pp 173–181, 2014.

[IYE 02] IYER A.K., ELEFTHERIADES G.V., "Negative refractive index meta materials supporting 2-D waves", in *IEEE MZ7'-S International Microwave Symposium Digest*, vol. 2, pp. 1067–1070, Seattle, WA, June 2002.

[IYE 03] IYER A.K., GRBIC A., ELEFTHERIADES G.V., "Sub-wavelength focusing inloaded transmission line negative refractive index meta materials", in *IEEE MTTS International Microwave Symposium Digest*, Philadelphia, PA, pp. 199–202, June, 2003.

[IYE 03] IYER A.K., KREMER P.C., ELEFTHERIADES G.V., "Experimental and theoretical verification of focusing in a large, periodically loaded transmission line negative refractive index meta material", *Optics Express*, vol. 11, pp. 696–708, 2003.

[JIN 08] JIN H., WEN G., "A novel four-way Ka-band spatial power combiner based on HMSIW", *IEEE Microwave and Wireless Components Letters*, vol. 18, no. 8, pp. 515–517, 2008.

[KER 75] KERR A.R., "A technique for determining the local oscillator waveforms in a microwave mixer", *IEEE Transactions on Microwave Theory and Techniques*, vol. 23, pp. 828–831, 1975.

[KIM 06] KIM D.H., CHOI J.I., "Design of a multiband frequency selective surface", *ETRI Journal,* vol. 28, no. 04, pp. 506–508, 2006.

[KRU 94] KRUMPHOLZ M., RUSSER P., "A field theoretical derivation of TLM", *IEEE Transactions on Microwave Theory and Techniques*, vol. 42, Part 1–2, pp. 1660–1668, 1994.

[KUR 65] KUROKAWA K., "Power waves and the scattering matrix", *IEEE Transaction on Microwaves Theory and Techniques*, vol. 13, pp. 194–202, March 1965.

[LA 08] LA CRUZ W., RAYDAN M., "Residual iterative schemes for large-scale nonsymmetric positive definite linear systems", *Computational and Applied Mathematics*, vol. 27, no. 2, pp. 151–173, 2008.

[LUE 78] LUEBBERS R.J., MUNK B.A., "Some effect of dielectric loading on periodic slots arrays", *IEEE Transactions on Antennas and Propagation*, vol. 26, no. 4, pp. 536–542, 1978.

[MAC 05] MACCHIARELLA G., TAMIAZZO S., "Design techniques for dual-passband filters", *Transactions on Microwave Theory and Techniques*, vol. 53, no. 11, 3265–3271, 2005.

[MAR 01] MARTYNYUK A.E., MARTINEZ LOPEZ J.I., "Frequency selective surfaces based on shorted ring slots", *Electronics Letters*, vol. 37, no. 5, pp. 268–269, 2001

[MAR 05] MARTYNYUK A.E., MARTINEZ LOPEZ J.I., MARTYNYUK N.A., "Active frequency-selective surfaces based on loaded ring resonators", *Electronics Letters*, vol. 41, no. 1, pp. 2–4, 2005

[MIL 07] MILITARU N., LOJEWSKI G., BANCIU M.G., "Aperture couplings in multilayer filtering structures," *Signals, International Symposium on Circuits and Systems*, vol. 2, pp. 1–4, 2007.

[MON 05] MONACELLI B., PRYOR J.B., MUNK B.A. *et al.*, "Infrared frequency selective surface based on circuit-analog Square loop design," *IEEE Transactions on Antennas and Propagation,* vol. 53, no. 2, pp. 745–752, 2005.

[MON 07] MONNI S., GERINI G., NETO A. *et al.*, "Multimode equivalent networks for the design and analysis of frequency selective surfaces", *IEEE Transactions on Antennas and Propagation,* vol. 55, no. 10, pp. 2824–2835, 2007.

[PAN 03] PAN G.W., TRETIAKOV Y.V., GILBERT B. "Smooth local Cosine based Galerkin Method for Scattering Problems", *IEEE Transactions on Antennas and Propagation* , vol. 51, no. 6, 2003.

[PEN 00] PENDRY J.B., "Negative refraction makes a perfect lens", *Physical Review Letters*, vol. 85, no. 18, pp. 3966–3969, 2000.

[PEN 99] PENDRY J.B., HOLDEN A.J., ROBBINS D.J. *et al.*, "Magnetism from conductors and enhanced nonlinear phenomena", *IEEE Transactions on Microwave Theory and Techniques*, vol. 47, no. 11, pp. 2075–2084, November 1999.

[PIL 95] PILOTO A.J., LEAHY K.A., FLANICK B.A. *et al.*, Waveguide filters having a layered dielectric structures, US patent 5382931A, January 1995.

[QUI 06] QUING A., "Vector spectral-domain method for the analysis of frequency selective surfaces", *Progress in Electromagnetics Research*, vol. 65, pp. 201–232, 2006.

[RAV 03a] RAVEU N., Contribution au développement de la formulation en ondes en coordonnés cylindriques, application à l'étude d'antennes placées sur un cylindre parfaitement conducteur, INPT Thesis, Toulouse, 2003.

[RAV 03b] RAVEU N., VUONG T.P., TERRASSE I. *et al.*, "Near fields evaluated with the wave concept iterative procedure method for an e-polarisation plane wave scattered by cylindrical strips", *Microwave and Optical Technology Letters*, vol. 38, no. 05, pp. 403–406, September 2003.

[RAV 04] RAVEU N., VUONG T.P., TERRASSE I. *et al.*, "Wave concept iterative procedure applied to cylinders", *IEEE Proceedings of Microwaves Antennas and Propagation*, vol. 151, no.5, pp. 409–416, October 2004.

[RAV 09] RAVEU N., BAUDRAND H., "Improvement of the WCIP convergence", *IEEE AP-S*, 2009.

[RAV 10] RAVEU N., GIRAUD L., BAUDRAND H., "WCIP acceleration", *APMC*, pp. 971–974, 2010

[RAV 11] RAVEU N., GIRAUD L., BAUDRAND H., "Accélération de la WCIP", *JNM*, 2011.

[RAV 12] RAVEU N., VINCENT J., POIRIER J-R., "Physically-based preconditioner for the WCIP", *APMC*, pp. 1310–1312, 2012.

[ROS 95] ROSS D.C., VOLAKIS J.L., ANASTASSIU H.T., "Overlapping modal and geometric symmetries for computing jet-engine inlet scattering", *IEEE Transactions on Antennas Propagation*, vol. 43, no. 10, pp. 1159–1163, October 1995

[SAA 86] SAAD Y., SCHULTZ M-H., "GMRES: a generalized minimal residual algorithm for solving non symmetric linear systems", *SIAM Journal of Science and Statics Computation*, vol. 7, no. 3, July 1986.

[SBO 01] SBOUI N., GHARSALLAH A., GHARBI A. *et al.*, "Global modelling of microwave active circuits by an efficient iterative procedure", *Microwaves, Antennas and Propagation, IEEE Proceedings*, vol. 148, no. 3, pp. 209–212, June 2001.

[SER 10] SERRES A., FONTGALLAND G., DE FARIAS J.E.P. *et al.*, "An efficient algorithm for planar circuits design", *IEEE Transactions on Magnetics,* vol. 46, no. 8, pp. 3441–3444, 2010.

[SER 13] SERRES A., SERRES G.K.F., FONTGALLAND G. *et al.*, "Ideal polarizer using the WCIP", *29th International Review of Progress in Applied Computational Electromagnetics*, Monterey, pp. 1–4, 2013.

[SHI 05] SHIJO T., HARINO T. ANDO M., "Large–size local-domain basis functions with phase detour and fresnel zone threshold for sparse reaction matrix in the Method of Moments", *IEEE Transactions on Electron Devices, E88-C*, pp 2208–2215, December 2005.

[SHI 94] SHIGEKI F., Waveguide line, Japanese patent: JP 06 053711, February 1994.

[SIM 04] SIMPSON J.J., TAFLOVE A., MIX J.A. *et al.*, "Computational and experimental study of a microwave electromagnetic bandgap structure with waveguiding defect for potential use as a bandpass wireless interconnect", *IEEE Microwave and Wireless Components Letters*, vol. 14, no. 7, pp. 343–345, July 2004.

[SUT 03] SUTONO A., LASKAR J., SMITH W.R., "Design of miniature multilayer on-package integrated image-reject filters", *IEEE Transactions on Microwave Theory and Techniques*, vol. 51, no. 1, pp. 156–162, January 2003.

[TIT 07a] TITAOUINE M., GOMES NETO A., BAUDRAND H. *et al.*, "Analysis of frequency selective surface on isotropic/anisotropic layers using WCIP method", *ETRI Journal,* vol. 29, pp. 36–44, February 2007.

[TIT 07b] TITAOUINE M., GOMES NETO A., BAUDRAND H. *et al.*, "WCIP method applied to active frequency selective surfaces", *Journal of Microwave and Optoelectronics,* vol. 06, no. 01, pp. 1–16, June 2007.

[TIT 08a] TITAOUINE M., GOMES NETO A., BAUDRAND H. *et al.*, "Analysis of shorted ring slots frequency selective surfaces using WCIP method", *Journal of Microwaves, Optoelectronics and Electromagnetic Applications,* vol. 7, no. 2, pp. 65–83, December 2008.

[TIT 08b] TITAOUINE M., GOMES NETO A., BAUDRAND H. *et al.*, "Determination of metallic ring FSS scattering characteristics using WCIP method", *Microwave and Optical Technology Letters,* vol. 50, no. 05, pp. 1324–1328, May 2008.

[TIT 09] TITAOUINE M., RAVEU N, GOMES NETO A. *et al.*, "Electromagnetic modelling of quasi-square open metallic ring frequency selective surface using wave concept iterative procedure", *ETRI Journal*, vol. 31, no. 1, pp. 77–79, February 2009.

[TIT 10] TITAOUINE M., RAVEU N., GOMES NETO A. *et al.*, "Dual-band and enhanced band FSS characterization using WCIP method", *Microwave and Optical Technology Letters,* vol. 52, no. 5, pp. 836–839, April 2010.

[TIT 13] TITAOUINE M., RAVEU N., NETO A.G. *et al.*, "The WICIP method: theory and applications", in *Advances in Engineering Research*, Nova Publishers, USA, vol. 6, January 2013.

[TRA 03] TRABELSI H., GHARSALLAH A., BAUDRAND H. *et al.*, "Global electromagnetic analysis of microwave circuits including lumped elements based on iterative method", *Antennas and Propagation Society International Symposium, IEEE*, vol. 3, pp. 742–745, June 2003.

[TSA 07] TSAI C.-C., WU S.-T., "A new single surface integral equation for light scattering by circular dielectric cylinders", *Optics Communications*, vol. 277, pp. 247–250, 2007.

[UCA 08] UCAR M.H.B., SONDAS A., ERDEMLY Y.E., "Switchable split-ring frequency selective surfaces", *Progress in Electromagnetics Research B,* vol. 6, pp. 65–79, 2008.

[VOY 06] VOYER D., AUBERT H., DAVID J., "Scale changing technique for the electromagnetic modeling of planar self-similar structures", *IEEE Transactions on Antennas and Propagation,* vol. 54, no. 10, pp. 2783–2789, October 2006.

[WAH 92] WAHID M., MORRIS S.B., "Metal radomes reduced RCS performance", *GEC Journal of Research*, vol. 9, no. 3, pp. 166–171, 1992.

[WAN 03] WANE S., BAJON D., BAUDRAND H., "Full wave analysis of stacked structures including in homogeneous layers. Hybrid integral-differential approach in terms of waves", *Proc. New Trends and Concepts in Microwave Theory and Techniques Conf.*, Research Signpost, March 2003.

[WAN 05] WANE S., BAJON D., BAUDRAND H. *et al.*, "A new full-wave hybrid differential-integral approach for the investigation of multilayer structures including nonuniformly doped diffusions", *IEEE Transactions on Microwave and Technique*, vol. 53, no. 1, pp. 200–214, January 2005.

[WU 03] WU K., CASSIVI Y., "Low cost microwave oscillator using substrate integrated waveguide cavity", *IEEE Microwave and Wireless Components Letters*, vol. 13, no. 2, pp. 48–50, 2003.

[YAN 00] YANG H-Y.D., KIN R., JACKSON D.R., "Design consideration for modeless integrated circuit substrates using planar periodic patches", *IEEE Transactions on Microwave Theory and Techniques*, vol. 48, no. 12, pp. 2233–2239, December 2000.

[YAN 04] YAN L., HONG W., HUA G. *et al.*, "Simulation and experiment on SIW slot array antennas", *IEEE Microwave and Wireless Components Letters*, vol. 14, no. 9, pp. 446–448, sept. 2004.

[YIN 02] YIN W.-Y., LI L.-W., YEO T.-S. "The near-zone field characteristics of an E-polarization planewave penetrating through cylindrical multiple apertures (non) coated with lossy and lossless media", *IEEE Transactions Electromagnetic Compatation*, vol. 45, no. 2, pp. 329–337, May 2002.

[YON 81] YONEYAMA T., NISHIDA S., "Nonradiative dielectric waveguide for millimeter-wave integrated circuits", *IEEE Transaction and Microwave Theory Techniques*, vol. 29, no. 11, pp. 1188–1192, November 1981.

[ZAI 10] ZAIRI H., BAUDRAND H., GHARSALLAH A. *et al.*, "An efficient iterative method for analysis of a substrate integrated waveguide structures", *Microwave and Optical Technology Letters*, vol. 52, no. 1, pp. 45–48, January 2010.

[ZHO 96] ZHOU L., DAVIS L.E., "Finite element method with edge elements for waveguides loaded with ferrite magnetized in arbitrary direction" *IEEE Transactions on Microwave Theory and Techniques*, vol. 44, no. 6, pp. 809–815, June 1996.

List of Authors

Taoufik AGUILI
Syscom ENIT
Tunisia

Noemen AMMAR
Syscom ENIT
Tunisia

Med Karim AZIZI
Tunis el Manar University
Tunisia

Henri BAUDRAND
Laboratoire Laplace
INP Toulouse
France

Ahmad ISMAIL ALHZZOURY
INTP, UPS, LAPLACE,
ENSEEIHT, CNRS
University of Toulouse
France

Nathalie RAVEU
INTP, UPS, LAPLACE,
ENSEEIHT, CNRS
University of Toulouse
France

Alexandre Jean René SERRES
Federal University of Campina Grande
Brazil

Georgina Karla DE FREITAS SERRES
Federal University of Campina Grande
Brazil

Mohammed TITAOUINE
University of Batna 2
Algeria

Index

A, C, D

E, F, H

I, M, N

P, Q, S

T, W

Other titles from

iSTE

in

Waves

2016

ANSELMET Fabien, MATTEI Pierre-Olivier
Acoustics, Aeroacoustics and Vibrations

PARET Dominique
Antennas Designs for NFC Devices

PARET Dominique
Design Constraints for NFC Devices

WIART Joe
Radio-Frequency Human Exposure Assessment

2015

PICART Pascal
New Techniques in Digital Holography

2014

APPRIOU Alain
Uncertainty Theories and Multisensor Data Fusion

JARRY Pierre, BENEAT Jacques N.
RF and Microwave Electromagnetism

LAHEURTE Jean-Marc
UHF RFID Technologies for Identification and Traceability

SAVAUX Vincent, LOUËT Yves
MMSE-based Algorithm for Joint Signal Detection, Channel and Noise Variance Estimation for OFDM Systems

THOMAS Jean-Hugh, YAAKOUBI Nourdin
New Sensors and Processing Chain

TING Michael
Molecular Imaging in Nano MRI

VALIÈRE Jean-Christophe
Acoustic Particle Velocity Measurements using Laser: Principles, Signal Processing and Applications

VANBÉSIEN Olivier, CENTENO Emmanuel
Dispersion Engineering for Integrated Nanophotonics

2013

BENMAMMAR Badr, AMRAOUI Asma
Radio Resource Allocation and Dynamic Spectrum Access

BOURLIER Christophe, PINEL Nicolas, KUBICKÉ Gildas
Method of Moments for 2D Scattering Problems: Basic Concepts and Applications

GOURE Jean-Pierre
Optics in Instruments: Applications in Biology and Medicine

LAZAROV Andon, KOSTADINOV Todor Pavlov
Bistatic SAR/GISAR/FISAR Theory Algorithms and Program Implementation

LHEURETTE Eric
Metamaterials and Wave Control

PINEL Nicolas, BOURLIER Christophe
Electromagnetic Wave Scattering from Random Rough Surfaces: Asymptotic models

SHINOHARA Naoki
Wireless Power Transfer via Radiowaves

TERRE Michel, PISCHELLA Mylène, VIVIER Emmanuelle
Wireless Telecommunication Systems

2012

LALAUZE René
Chemical Sensors and Biosensors

LE MENN Marc
Instrumentation and Metrology in Oceanography

LI Jun-chang, PICART Pascal
Digital Holography

2011

BECHERRAWY Tamer
Mechanical and Electromagnetic Vibrations and Waves

GOURE Jean-Pierre
Optics in Instruments

GROUS Ammar
Applied Metrology for Manufacturing Engineering

LE CHEVALIER François, LESSELIER Dominique, STARAJ Robert
Non-standard Antennas

2010

BEGAUD Xavier
Ultra Wide Band Antennas

MARAGE Jean-Paul, MORI Yvon
Sonar and Underwater Acoustics

2009

BOUDRIOUA Azzedine
Photonic Waveguides

BRUNEAU Michel, POTEL Catherine
Materials and Acoustics Handbook

DE FORNEL Frederique, FAVENNEC Pierre-Noël
Measurements using Optic and RF Waves

FRENCH COLLEGE OF METROLOGY
Transverse Disciplines in Metrology

2008

FILIPPI Paul J.T.
Vibrations and Acoustic Radiation of Thin Structures

LALAUZE René
Physical Chemistry of Solid-Gas Interfaces

2007

KUNDU Tribikram
Advanced Ultrasonic Methods for Material and Structure Inspection

PLACKO Dominique
Fundamentals of Instrumentation and Measurement

RIPKA Pavel, TIPEK Alois
Modern Sensors Handbook

2006

BALAGEAS Daniel *et al.*
Structural Health Monitoring

BOUCHET Olivier *et al.*
Free-Space Optics

BRUNEAU Michel, SCELO Thomas
Fundamentals of Acoustics

FRENCH COLLEGE OF METROLOGY
Metrology in Industry

GUILLAUME Philippe
Music and Acoustics

GUYADER Jean-Louis
Vibration in Continuous Media

9 781848 219595

Printed at Repro India Ltd.